ICT and Rural Development in the Global South

This book dives into the achievements, opportunities, risks and dangers of ICT in the rural Global South, and takes a look at the likely future.

Drawing on years of experience across 45 counties, as well as extensive original academic research, Willem van Eekelen situates the evolving role of ICT in wider development patterns in the Global South. He discusses the effects of ICT on agriculture, trade, financial flows, resource management and governmental performance. He then considers the associated risks of financial insecurity, online gambling, exclusion, misinformation and the effects of ICT on people's freedom. The book concludes with six recommendations to maximise the usefulness of rural ICT investments and minimise the risk of them causing harm.

This engaging and authoritative account of ICT and rural development will help students, academics, governmental policymakers, donors and investors wishing to support socio-economic development in the Global South.

Willem van Eekelen is an economist who worked for a range of companies, donors, UN agencies and non-governmental organisations – always in the broad field of international development. Until 2010, Willem mostly held policy and programme positions. Since then, he has worked as an independent evaluator in the humanitarian and development sectors. In parallel, Willem supports organisations with their strategy development, holds occasional trusteeships and teaches at the University of Birmingham. Willem co-owns Green Visions, a rural adventure tourism company. His previous publications include academic works (including Routledge's *Rural Development in Practice*) as well as a few tourist guides and Bloomsbury's *100 ideas for Dads who love their kids but find them exhausting*.

Rethinking Development

Rethinking Development offers accessible and thought-provoking overviews of contemporary topics in international development and aid. Providing original empirical and analytical insights, the books in this series push thinking in new directions by challenging current conceptualizations and developing new ones.

This is a dynamic and inspiring series for all those engaged with today's debates surrounding development issues, whether they be students, scholars, policy makers and practitioners internationally. These interdisciplinary books provide an invaluable resource for discussion in advanced undergraduate and postgraduate courses in development studies as well as in anthropology, economics, politics, geography, media studies and sociology.

Researching Development NGOs
Global and Grassroots Perspectives
Edited by Susannah Pickering-Saqqa

South-North Dialogues on Democracy, Development and Sustainability
Edited by Cristina Fróes de Borja Reis and Tatiana Berringer

Dear Development Practitioner: Advice for the Next Generation
Edited by Simon Milligan and Lee Wilson

Foreign Aid and its Unintended Consequences
Dirk-Jan Koch

Reformers in International Development
Five Remarkable Lives
David M. de Ferranti

Young People in the Global South
Voice, Agency and Citizenship
Edited by Kate Pincock, Nicola Jones, Lorraine Van Blerk and Nyaradzayi Gumbonzvanda

For more information about this series, please visit: www.routledge.com/Rethinking-Development/book-series/RDVPT

ICT and Rural Development in the Global South

Willem van Eekelen

Routledge
Taylor & Francis Group

LONDON AND NEW YORK

Illustrations, including cover image: Micky Dirkzwager

First published 2024
by Routledge
4 Park Square, Milton Park, Abingdon, Oxon OX14 4RN

and by Routledge
605 Third Avenue, New York, NY 10158

Routledge is an imprint of the Taylor & Francis Group, an informa business

© 2024 Willem van Eekelen

British Library Cataloguing in Publication Data
A catalogue record for this book is available from the British Library

Library of Congress Cataloging-in-Publication Data
A catalog record has been requested for this book

ISBN: 9781032588445 (hbk)
ISBN: 9781032588421 (pbk)
ISBN: 9781003451716 (ebk)

DOI: 10.4324/9781003451716

Typeset in Sabon
by Taylor & Francis Books

Contents

Table and boxes

Table

Boxes

Abbreviations and acronyms

2G / 3G / 4G/ 5G	$2^{nd}/3^{rd}/4^{th}/5^{th}$-generation cellular network
$2G^{+}$	2G and above (2G means 2nd-generation cellular network)
ADB	Asian Development Bank
AfBD	African Development Bank
AFC	Agence Française de Développement
AFIC	Africa Freedom of Information Center
AIDS	Acquired immunodeficiency syndrome
AiTM	Adversary-in-the-middle (i.e., a third party that inserted itself into people's virtual communication to overhear or manipulate it)
AML	Anti-money laundering
APG	Algemene Pensioen Groep
API	Application Programming Interface
ATM	Automated teller machine
BBC	British Broadcasting Corporation
BiH	Bosnia and Herzegovina
BRAC	Building Resources Across Communities (used to be Bangladesh Rural Advancement Committee)
CBK	Central Bank of Kenya
CBNRM	Community-based natural resource management
CCTA	Computer and Computing Technologies in Agriculture
CFT	Combating the financing of terrorism
CGIAR	Consultative Group for International Agricultural Research
CIC	Community Information Centre
CIJM	Centre for Investigative Journalism Malawi
CIMMYT	International Maize and Wheat Improvement Centre

CIPS	Citizen Identification Protection System
CIW	China Internet Watch
COP27	27th United Nations Climate Change conference, held in Sharm El Sheikh in Egypt in November 2022
Covid-19	Coronavirus disease, caused by the SARS-CoV-2 virus
CSEIFG	Capacity Strengthening and Empowerment to Improve Forest Governance
CSO	Civil Society Organisation (this book uses CSO and NGO interchangeably)
CSR	Corporate Social Responsibility
DESA	United Nations Department of Economic and Social Affairs
DFC	US International Development Finance Corporation (formerly OPIC, the Overseas Private Investment Corporation)
DFID	Department for International Development (UK), nowadays part of FCDO, the UK's Foreign, Commonwealth & Development Office
DGGF	Dutch Good Growth Fund
E-KYC	Digital 'know-your-customer'
FAO	Food and Agriculture Organization
FATF	Financial Action Task Force
FB	Facebook
FCDO	Foreign, Commonwealth & Development Office (UK)
FMO	Entrepreneurial Development Bank
FSD Kenya	Financial Sector Deepening Kenya
G-CSPI	Global Correlation Sensitive Poverty Index
G8	Group of eight countries (Canada, France, Germany, Italy, Japan, Russia, United Kingdom, United States)
GDP	Gross domestic project
GIZ	Deutsche Gesellschaft für Internationale Zusammenarbeit
GNI	Gross national income
GPS	Global Positioning System
HIV	Human immunodeficiency virus
IBAI	Index-based agricultural insurance
ICAI	Independent Commission for Aid Impact
ICT	Information and communication technology

ICT4D	ICT for development / Information and communication technology for development
ID	Identity document
IFC	International Finance Corporation
ILO	International Labour Organization
IPCC	Intergovernmental Panel on Climate Change
ITU	International Telecommunication Union
JCM	The John C. Martin Foundation
JPO	Junior Professional Officer
KfW	German Development Bank
KNAW	Koninklijke Nederlandse Akademie van Wetenschappen
KNBS	Kenya National Bureau of Statistics
KYC	Know-your-customer
LAN	Local area network
LGBT+	LGBT refers to lesbian, gay, bisexual and transgender. The plus refers to other sexual minorities.
M4P	Making markets work for [the] poor
MIS	Management Information System
MMW4P	Making markets work for [the] poor
MOFCOM	Ministry of Commerce of the People's Republic of China
MRA	Malawi Revenue Authority
MTO	Money transfer operator
NBER	National Bureau of Economic Research (USA)
NGO	Non-governmental organisation (this book uses NGO and CSO interchangeably)
Norad	Norwegian Agency for Development Cooperation
NWO	Netherlands Organisation for Scientific Research
ODA	Official Development Assistance, which represents flows of official financing administered with the promotion of the economic development and welfare of developing countries as the main objective
ODK	OpenDataKit
OECD	Organisation for Economic Co-operation and Development
OHR	Office of the High Representative (for Bosnia and Herzegovina)
OPIC	Overseas Private Investment Corporation (nowadays DFC, the US International Development Finance Corporation)

PACM	Proceedings of the Association for Computing Machinery
PIC	Peace Implementation Council (for Bosnia and Herzegovina)
PPA	Programme Partnership Arrangement
PR	Public relations
PUS	Public Understanding of Science
QR	Quick response
R&D	Research and development
REDD+	REDD+ is a climate change mitigation approach. REDD stands for the Reduction of Emissions from Deforestation and forest Degradation, and the '+' refers to the focus on fostering conservation, sustainable management of forests, and enhancement of forest carbon stocks.
RMB	Renminbi, China's official term for the yuan (with subtle differences that do not matter for this book)
RML	Reuters Market Light (full name no longer in use)
RNFE	Rural non-farm economy
RRI	Responsible research and innovation
RVO	Netherlands Enterprise Agency
SAFG	Strengthening African Forest Governance
SDG	Sustainable Development Goal
SIG	Sport Industry Group
SIM	Subscriber identification module
SMS	Short message service
TAM	Technology Acceptance Model
Triple-F	Food, Fuel, Finance
UK	United Kingdom
UN	United Nations
UNEP	United Nations Environment Programme
UNESCO	United Nations Educational, Scientific and Cultural Organization
UNGA	United Nations General Assembly
UNHCR	The UN Refugee Agency
USA	United States of America
USAID	US Agency for International Development
UTAUT	Unified Theory of Acceptance and Use of Technology
VPN	Virtual Private Network
VSLA	Village Saving and Loan Association
VSO	Voluntary Services Overseas

WED	Women's entrepreneurship development
WEE	Women economic empowerment
WEF	World Economic Forum
WHO	World Health Organization
WWF	World Wildlife Fund (now World Wide Fund for Nature)

Acknowledgements

The biggest perk of conducting evaluations (which is what I do for a living) is that I get to listen to rural women, men and children from all over the world. I am grateful to all who shared their thoughts and experiences with me. I am also grateful to the many development professionals I interviewed as part of my work. In no more than an hour they shared insights gained in the course of years, and this often made for steep learning indeed.

I'd like to thank Paula Bownas for fine-tuning some of the text; Micky Dirkzwager for the illustrations (aren't they great?!); and Helena Hurd and Katerina Lade for the enthusiasm and speed with which they turned this text into a book.

I would like to thank Maha, my dear wife, for being my dear wife. We both work internationally, so we don't see each other a whole lot, but when we're together she's the best possible reason for taking a break.

Prologue

My name is Willem van Eekelen and I work as an independent evaluator of development efforts in the Global South. Many of my evaluations include projects that use ICT with the aim of enhancing rural socio-economic development. In this prologue I look back at a few of these projects, and describe how my early enthusiasm for 'ICT for Development' (ICT4D) cooled down and lost its naivety, but did not disappear altogether, and how I came to the issues I seek to cover in this book. This prologue focuses on education but could have focused on almost any other sector as ICT affects every part of rural life.

—

Some 20 or so rural Malawian kids were sitting on the classroom floor. Each one had a tablet and the exercises they were doing were meant to strengthen their numerical skills. Their 'edutainment' required no literacy: it was all picture-based, with a woman's voice giving easy instructions in the local language. Some of the children recognised that six corncobs fit on six plates. Others tried it with five or seven. One or two merely moved the corncobs around, patiently but without evidence of learning or sense of enjoyment.

I was standing there, as the evaluator of this pilot programme, and made my observations with a 'business as usual' attitude. The notes I took at the time are a little dry:

- A few kids get it and like it.
- The teacher engages equally with girls and boys...
- ... but she ignores the kids with learning difficulties, as well as the girl who seems capable of doing far more challenging exercises.
- Two of the tablets do not work. A third tablet is about to break down as the child is poking it too hard.

DOI: 10.4324/9781003451716-1

I was getting bored. This was the second school I had visited and I thought the strengths and weaknesses of this little pilot were obvious.

Then I looked up, and out of the window. Out in the open, a large group of kids was sitting under the sun, with a single suit-wearing teacher standing in the shade of a solitary tree, legs apart, surrounded by his pupils. Some were trying to get their exercise books into his hands. He took them without eye contact, signed them without looking, and threw them back. Others were just sitting there, waiting for the school day to be over.

This woke me up and triggered the techno-optimist in me. Suddenly, the tablets were fascinating, as *they had the potential to bypass problems that decades of education strategies, teacher training, curriculum development and foreign support had not managed to address.* Even in large classrooms, tablets would allow children to become their best selves, at their own pace. Many of them would be able to do this without the support of a qualified, motivated teacher. Kids would *enjoy* working on these tablets, as e-lessons were more interesting, challenging and interactive than traditional teaching. This might keep them in school for longer, and help their speed of learning. The teachers' login, in the morning, could be used to monitor the massive problem of teacher absenteeism (and in fact the Head of the school felt this was the tablets' most important benefit). I was sold. These tablets[1] had the potential of achieving a step-change in the learning of millions of rural kids.

A few days later, I flew from Malawi to Papua New Guinea, as part of an assessment of the effectiveness of that same international non-governmental organisation (NGO),[2] and saw the pilot stage of a programme that sent daily text messages to teachers in remote areas, with lesson plans and exercises they could use during that day. An assessment of the results after 100 school days of this pilot concluded that kids had more chance of progressing their learning if they were taught by teachers who received these daily text messages than if they were taught by teachers who did not.[3] I returned home convinced that ICT was going to make a massive difference to education in even the world's most remote schools.

This was a decade ago. In the years since, I have often been in awe of the way ICT innovations powered my own learning and teaching. Nowadays, online texts in all major languages translate into English with the click of a button. Almost every useful paper exists in electronic format and is easy to find. Every year my university offers new virtual tools enabling students to engage with each other and me with evermore ease. My face-to-face continuous professional development investments are few and far between, but digital options are so plentiful, appealing, imposing and instantly accessible that they sometimes distract me from doing actual work. Their advanced features often incur a fee, but the basic versions of virtual communication software applications are open-source and many learning products are open-access. This avalanche of new possibilities shows no sign of abating, and the Covid-19 pandemic even accelerated investments in remote learning products that are now starting to spill over to poorer and as yet unconnected rural regions. Many such products focus on personalised learning and they may increasingly outcompete formal, traditional education, because they are more versatile and potentially more adaptable to an individual's needs than a classroom teacher – and because they are heavily promoted by the companies that invented them.

Overall, however, I have seen too many defunct and deserted rural school computer rooms, and too many other ICT failures, to maintain my overall sense of enthusiasm. As it turns out, ICT is not the silver bullet for rural learning I briefly and naively believed it to be. Even projects with promising starts often crumble a few years later, when the newness has faded and donors and implementers have left without the products and services they introduced having been truly absorbed by the local educational system. That text message initiative in Papua New Guinea, for example, was full of promise during its early stages of implementation... and then it collapsed. In my 2020 book on rural development, I wrote enthusiastically about the high-profile and award-winning 'hole-in-the-wall' concept of freely available tablets, embedded in public walls, that kids in mostly Asian countries (and Uganda) use to learn and play games.[4] However, Payal Arora had "encountered Mitra's Hole-in-the-Wall experiment, not as a hub of thriving autonomous learning but instead as abandoned gaping holes, forgotten by the passer-by".[5] Far more costly endeavours also sometimes fail to result in meaningful benefits. For example, the One Laptop per Child pro-gramme that distributed over 2 million laptops in some 40 countries in the Global South, at US$200 per laptop, was evaluated by using data from over 300 schools in rural Peru. This evaluation concluded that it had not increased kids' motivation to learn or to dedicate more time to homework; had not led to more reading (even though these laptops are loaded with 200 books); and had not enhanced the quality of school lessons. The overall conclusion was that:

> The program [led to] substantial increases in use of computers both at school and at home [but] no evidence is found of effects on test scores in math and language. There is some evidence, though inconclusive, about positive effects on general cognitive skills.[6]

Even significant positive effects do not necessarily mean that the ICT investments represent good value for money, as the benefits of alternative investments (that never took place and are therefore invisible and undiscussed)[7] might have been higher. As Julián Cristia and colleagues conclude:

> Governments should consider alternative uses of public funds before implementing large-scale technology in education programs [as] recent work on this topic suggests that technology in education programs that prioritize hardware provision are unlikely to provide a cost-effective and simple solution to educational disparities both across and within countries. In particular, in poor countries where teachers' salaries are low, the opportunity costs of implementing (capital-intensive) technology programs may be substantial compared with alternative labour-intensive education interventions, including reductions in class size and professional development.[8]

A frequent finding of assessments of educational ICT endeavours in the Global South is that ICT can be a force for good but that "the effective use of ICT [in education] is harder than was initially expected".[9] In part this is because the aid community's notion of 'effective use' is based on the implicit and erroneous assumption that poor people will use ICT applications for the productive purposes they were intended for. This is often untrue: ICT applications, including those that are meant to further education or health or economic development, are often most used for recreational purposes. Like children from wealthier backgrounds, "children living in poverty might watch pornography through government-gifted laptops and delete their homework to create space for their favorite music downloads".[10]

A similar prologue is possible for other fields where ICT is influencing rural life. This is because ICT influences all parts of it, and in each of these parts it may sometimes have transformative *potential* but often fails to achieve positive (defined as 'development-relevant') effects. Moreover, some ICT applications, and the overall connectivity of people, come with fundamental risks and dangers. Some of them have already materialised, with disastrous consequences.

This book explores ICT's potential benefits, challenges and risks, and the damage it is causing, in relation to inclusive and sustainable socioeconomic development in the rural Global South. It seeks to make a twofold contribution to the literature on ICT in the context of rural development.

1   I use my evaluative data and experience and recent empirical literature (and recency is important as this field is evolving very rapidly) to explore and assess a wide range of ways in which ICT may be benefiting rural development throughout the Global South. Examples of potential benefits are efficiency gains in production, trade and the financial sector; more stable, predictable and higher levels of income for poorer segments in rural society in particular; more efficient and less corrupt governance; and the reduction of greenhouse gas emissions and the conservation of nature. I am not aware of other publications that do this, with a focus on the rural regions within the Global South and without implicitly pursuing a pro-ICT agenda.[11]

2   I look at early trends, positive and negative, and guesstimate how these trends are likely to develop. In addition, I use retroduction[12] to identify ICT-related opportunities, risks and dangers that are not rooted in current trends and are not yet covered in existing ICT literature. The application of retroduction serves as a type of horizon scanning for unintended effects of ICT innovations – and because ICT developments and their utilisation evolve so very rapidly, these horizons may not be many years away.

This book consists of two parts. The first part presents seven broad fields in which ICT applications affect rural life in the Global South. These are agricultural production, trade, livelihood diversification, financial flows, e-government, climate change, and natural resource management. The final chapter of this first part explores how and why some ICT applications have spread at such speed. The second part of the book explores the risks and dangers of the rapid spread of ICT, for business and pleasure, in rural life. This part explores ICT's effects on farming and rural employment, financial safety and security, gambling, inequalities, misinformation and civic space. The book ends with a set of conclusions, and with a few reflections on the ways in which donors and angel investors[13] could finance ICT products and services to enhance socio-economic development, whilst mitigating the risks they pose.

Notes

1   This was a pilot of a social company called onebillion (onebillion.org), implemented in cooperation with Voluntary Services Overseas (VSO).
2   This was the mid-term evaluation of a 'PPA' ('Programme Partnership Arrangement'). PPAs no longer exist but used to be highly sought-after, as

they were sizeable and more or less unrestricted block grants, provided by
DFID (the examples from this prologue are from PPAs with a total value of
$140 million). The unrestricted nature of the grants meant that PPA eva-
luations were not programme or country portfolio evaluations, but overall
organisational assessments that, among other things, led me to visit a range
of more or less randomly selected programmes in a range of countries.

3 Kaleebu, N. *et al* (2013) *SMS story impact assessment report*, VSO, Papua
New Guinea.

4 See, e.g., Mitra, S. and Dangwal, R. (September 2017) "Acquisition of computer literacy skills through self-organizing systems of learning among children in Bhutan and India", *Prospects*, volume 47, issue 3, pages 275–292. I did
not realise, at the time of writing the book, that this and indeed most publications on the hole-in-the-wall concept had been written by either the
inventor of the concept himself (Sugata Mitra), or by others from the company he works for – an observation made on page 107 of Arora, P. (2019)
The next billion users: digital life beyond the West, Harvard University Press.

5 *Ibid*, with the quotation from page 103 and a criticism of the PR around the
concept in the rest of the chapter. See also Arora, P. (2010) "Hope in the
Wall? A digital promise for free learning", *British Journal of Educational
Technology*, volume 41, issue 5, pages 689–702.

6 Cristia, J.P. *et al* (2017) "Technology and child development: evidence from
the one laptop per child program", *American Economic Journal: Applied
Economics*, volume 9, number 3, pages 295–320, with the quotation from
the abstract.

7 Nassim Taleb covers the issue of the popularity bias towards investments
that took place at the cost of invisible alternatives (as well as the bias
towards *responding to* problems at the cost of *preventing* problems) in
Taleb, N.N. (2008) *The Black Swan; the impact of the highly improbable*,
Penguin, Kindle Edition, especially pages 110–112.

8 Cristia, J.P. *et al* (2017) "Technology and child development: evidence from
the one laptop per child program", *American Economic Journal: Applied
Economics*, volume 9, number 3, pages 295–320, with the quotation on page
318.

9 Saïd, A., El Amrani, R. and Watson, R.T. (2010) "ICT and education: a
critical role in human and social development", *Information Technology for
Development*, volume 16, issue 3, pages 151–158, with the quotation from
page 152. This is the editorial introduction of an "ICT Education in
Development" special issue of *Information Technology for Development*.

10 Arora, P. (2019) *The next billion users: digital life beyond the West*,
Harvard University Press, with the quotation from page 4.

11 The World Bank's *World development reports* of 2016 and 2021 combine to
provide a wide-ranging overview, but these are publications with a pro-ICT
bias. See World Bank (2016) *World development report: digital dividends*, A
World Bank Group Flagship Report; and World Bank (2021) *World development report: data for better lives*, A World Bank Group Flagship Report.
The work of Mark Graham does not have a notable bias and he provides a
floor to both ICT optimists and ICT sceptics alike in, e.g., Graham, M.,
editor (2017) *Digital economies at global margins*, The MIT Press (which
does not have an exclusively rural focus but which does include several rural
case studies).

12 Retroduction is "non-valid logical reasoning where inferences are based on concomitance, on co-occurrences or striking similarities in behavioural patterns between the situation at hand and other, comparable situations". Tromp, C. (2017) *Wicked philosophy; philosophy of science and vision development for complex problems*, Amsterdam University Press, with the quotation from page 188. See Annex 1 for the application of retroduction in this book.

13 Angel investors finance the development of early-stage companies in return for part-ownership and may do so with both profit and charitable motives.

Part I

ICT's contributions to rural development

ICT is the sum total of technologies that are used to facilitate communication or to store, retrieve, use or manipulate information. Radio, television, computers and phones and their hardware and software, services such as internet kiosks, videoconferencing facilities, satellites: they and their many combinations all count as 'Information and Communication Technology'.

There are two very broad paradigms within which researchers interpret ICT, its effects and its dangers. First, there are techno-optimists, exemplified by ecomodernists and multilateral organisations such as the World Bank and UNEP. They want to 'dematerialise' (get ever-more outputs per unit of energy, land or other type of input) and eventually 'decouple' production from the harm it has historically caused the climate, nature and biodiversity. They believe that ICT could contribute to both aims, whilst simultaneously contributing to ongoing economic growth and socio-economic development. The large-scale agro-industry expresses a similar faith:

> The world's population is growing, but the amount of farmland available per head is shrinking. Agricultural productivity will have to increase if we want to safeguard our food supply in the long term. Digitalization in farming can help us deploy our resources efficiently and sustainably, enabling farmers to get the best out of their fields with minimal environmental impact.[1]

Techno-optimists are not normally majorly concerned about the risks, dangers and unpredictability of the effects of technological innovation.

Second, there are degrowth scholars. Degrowth scholars argue that "'sustainable growth' is a contradiction in terms [as] nothing physical can grow indefinitely."[2] They speak of a 'development paradox' (or the 'impossibility theorem') because "the environment is not equipped to absorb its unrelenting exploitation by the current growth model of

DOI: 10.4324/9781003451716-2

endless accumulation. In other words, development as we know it is undermining itself."[3] Some such scholars see good ICT initiatives even in today's world, such as Fairphone,[4] but most believe that "modern technology, in the presence of continued economic growth, does not promote sustainability but instead hastens growth".[5] ICT, and technology more broadly, can therefore not be a force for good, unless it is first 'appropriated' for the cause of degrowth. Until then, technology's effects are likely to be negative, and likely to further increase the power and wealth of a small group of large transnational companies.

A range of researchers and activists who do not self-identify as degrowth scholars or supporters do share some of their concerns. However, the donor community does not. Instead, or at least in my experience, donors nearly always side squarely with the techno-optimists.

Two opposing views on ICT

Interviews with donors suggest that they are enthusiastically investing in ICT because, after a series of demoralising setbacks that long preceded Covid-19's destruction of development progress, they are eager to find a new big winning formula and think that investing in rural ICT might be it. In the early 1990s, donors had had a brief moment of euphoria. The Berlin wall had fallen, and Fukuyama had posited, in *The end of history and the last man*,[6] that socio-economic development was now going to be a largely technical matter that would take place in a world that was moving ineluctably towards a global liberal democracy. This prediction proved to be untrue, and in the decades that followed donors have been hit by a number of disappointments.

The damage caused by structural adjustment programmes they had financed had become increasingly obvious. These programmes had discouraged and dismantled rural investments – especially ones that were not 'market conform' – and by 2008 even the World Bank itself admitted that these programmes had had seriously adverse consequences for many smallholders, and that "there is now general agreement that the state must invest in core public goods, such as agricultural R&D, rural roads, property rights, and the enforcement of rules and contracts."[7] The results of aid spending on governance, anti-corruption and democratisation programming had proven to be disappointing, and the Arab Spring had ended disastrously. Climate change predictions got progressively worse[8] and the November 2022 COP27 is but the latest example of countries failing to formulate game-changing commitments and follow up on them. Peacebuilding programmes could be individually successful but collectively failed to stop the trend of regress into ever-more widespread fragility,[9] and literature on how to use aid to reduce conflict and instability does not yet present much evidence on what has proven to work. The swiftness of the Taliban victory in Afghanistan, in August 2021, is a stark illustration of the ample but unsuccessful investments in peacebuilding. Even seemingly technical programmes with hugely promising prospects proved disappointing in reality. Microfinance has not been the success donors thought it was, for example, and the Green Revolution did not take root in Africa. In fact, it has become *bon ton*, in part of the popular press and academia, to be *so* dismissive of the entire global investment in 'Official Development Assistance' (ODA) that statements along the lines of "development aid [...] is failing"[10] no longer warrant substantiation or referencing.

The implication was that donors craved new types of programmes that promised a step change in socio-economic development of the Global South. The notion that ICT could have positive and indeed *transformative* effects on socio-economic development, and the idea that it could help reduce the harm humankind is causing to nature and climate, have therefore been easy sells. NGOs and the private sector recognised this, and developed proposals for ICT solutions, both as stand-alone projects and as add-ons that increased the appeal of conventional projects in agriculture and all other sectors in the broad field of rural development.

Many of these proposals fit all donor requirements. They were new and exciting and their usefulness had not yet been disproven. They supported the things donors felt strongly about. They would enhance food production and food security, through apps that supported precision farming. They were aligned with and supported free markets, through marketing and pricing apps. They would contribute to a thriving and inclusive financial sector, through new forms of microfinance. They would strengthen good governance, through e-government and social accountability and anti-corruption platforms. And they would enhance the position of women, through ICT solutions that make life easier *within* existing power relations and ones that *transform* these relations. They would help alleviate extreme poverty through social assistance programmes that were much-improved after their transition to virtual payments. All these were issues that had taken a hit after the unanticipated triple-F crisis of 2008–09.[11] Now they needed rescuing, and ICT was just the thing to use for that. Moreover, ICT investments sometimes managed to achieve a lot in very little time. Mobile money took flight soon after it was introduced, mobile telephony leapfrogged over landlines, and every month new text- and smartphone-based products and services appeared on the market. Lastly, donor governments had responded to increasing xenophobia at home by promising that their ODA would tackle "the root causes of mass migration"[12] and, in this context, they thought that ICT-powered efforts to improve young people's livelihoods in 'migration countries' would reduce their desire to come to Europe.[13]

So donors embraced and funded ICT initiatives. Some NGOs and companies that developed rural-focused ICT products and services managed to attract *dozens* of grants (see Box 0.1). Donors often argue that the need for their support is temporary, because some of the ICT products and services are showing proof of concept in terms of commercial viability. This may soon achieve a demonstration effect (i.e., "look, such endeavours can be very profitable!") that will convince fully commercial financers to get involved even without junior debt reducing the risks they face. For example, a company named Babban Gona works with farmers in northern Nigeria – a very tough context to work in. At the time of writing, 30% of Babban Gona's investment capital is still concessional (to reduce the cost of capital), or junior (to derisk private investors and thereby have a catalysing effect), or both; but after only a few years of operations, Babban Gona's cumulative farmer repayment rates are already in excess of 98.5%.[14] This suggests that this type of more or less comprehensive ICT-based farm improvement functionality can indeed be profitable – even in the tough conditions of northern Nigeria. Donors assume that such successes become commonplace. The notion of the demonstration effect is a key part of the Theory of Change of the Dutch Good Growth Fund (DGGF),[15] for example, and a European donor's draft vision for 2030 is that, by 2030, "parts of the inclusive (and often digital) offer of the financial sector [have] reached tipping points of commercial viability".[16]

Box 0.1 Donors like ICT-powered socio-economic development initiatives – a *lot*

Most of the products and services covered in this book are not yet fully commercial. In many cases, I came across them because I make a living evaluating donor portfolios – and such portfolios tend to include lots of grants, concessional loans and other forms of support to NGOs and companies that produce ICT products and services that aim to strengthen socio-economic development. Almost invariably, I found these NGOs and companies to be supported by quite a range of donors. PharmAccess, for example, received grants and concessional funding from the Achmea Foundation, ADB, AFC, APG, DFC (called OPIC at the time of funding), DFID (now FCDO), FMO, the Gates Foundation, GIZ, Grand Challenges Canada, IFC, ILO, JCM, KfW, the Dutch Ministry of Foreign Affairs, the M-Pesa Foundation, MSD for mothers, Nationale Postcode Loterij, Norad, NWO, the Rockefeller Foundation, RVO, Saving Lives at Birth, USAID, World Bank, and several dozen donors and financers I did not recognise.[17] When I visited control group tech companies (i.e., ones that had not received concessional financing or other support from the donor I was evaluating) I found that they, too, had received grants, subsidised loans, junior equity financing and/or support from incubators and accelerators. I also visited a few of these incubators and accelerators – and they, too, were at least partly financed by institutional donors and foundations.

ODA even goes to very large companies that develop ICT products for the 'base of the pyramid' (originally 'bottom of the pyramid' – a business term that refers to the unexplored mass market of the world's poorest billions). MasterCard, for example, developed its 2KUZE app (which connects East African farmers, agents and buyers on a digital platform to optimise pricing transparency and distribution effectiveness) with financial support from the Gates Foundation[18] – one of many examples where ICT4D investments add to the concentration of market power.[19] Support to companies operating in the field of ICT4D does not merely come in the form of financing. There are many visibility-enhancing awards for ICT applications with developmental objectives as well (including from country governments, as a modern manifestation of the developmental state mentality), and many of the applications mentioned in this book have won one or more of them.

Donors often publicise ICT victories, sometimes on the basis of unverified result claims made by the companies behind the ICT products themselves, and highlight individual successes within programmes that did not in fact achieve significant overall results.[20] The grants I have seen often came with quite clear-cut targets and were provided in the expectation that technologies would achieve the effects they aimed to achieve. Donors do not pay much attention to ICT risks and uncertainties. They show little awareness of Kranzberg's First Law of Technology, which states that:

> Technology is neither good nor bad; nor is it neutral. [...] Technology's interaction with the social ecology is such that technical developments frequently have environmental, social, and human consequences that go far beyond the immediate purposes of the technical devices and practices themselves, and the same technology can have quite different results when introduced into different contexts or under different circumstances.[21]

This book gives many examples of donor investments that are underpinned by this optimism. The examples come in such numbers, in fact, that this book might create the impression that ICT applications are more widespread in the rural Global South than is actually the case. In reality, there are still plenty of regions where only the most basic ICT tools have gained traction.

In the context of rural development, four broad areas in which ICT applications could potentially amount to game changers stand out, and within these, I identify seven 'fields'.

(A)   **ICT could potentially enhance rural people's access to livelihood-related information.** This could possibly improve agricultural production processes (*Field 1*), contribute to a more equitable distribution of the benefits of trade (*Field 2*), and help rural households diversify their sources of income, including non-agricultural sources of income (*Field 3*).

(B) **ICT could potentially facilitate access to money.** It could possibly improve the reach and effectiveness of microfinance products and services (*Field 4*). It could also reduce the cost and enhance the reliability of the two main urban-to-rural flows of money that are not trade- or business-related: the remittances that individual migrants send 'back home', and state-led social assistance flows (both part of *Field 5*).

(C) **ICT could potentially enhance a state's functioning.** ICT could help governments improve their performance in some of the fields mentioned under A and B, such as in extension services and social assistance coverage. In addition and more generally, 'e-government' could possibly reduce bureaucracy and corruption (*Field 6*). Like the issues under area A above, this potential is grounded in the availability of information, this time not about agricultural processes, trade or income opportunities but about the state's systems, policies and processes.

(D) **ICT could potentially help mitigate against climate change and support natural resource management** (*Field 7*).

These seven fields are covered in the next seven chapters, after which Part I ends with an eighth chapter about the ways in which ICT products and services spread throughout the rural Global South. Part I is structured so that it very broadly moves from issues dominated by, respectively, access to production- and trade-related information, to access to finances, to state functioning, to environmental issues. However, the distinctions are not clear-cut and many ICT-driven companies cover more than one of these seven fields. The just-mentioned Babban Gona, for example, uses ICT as it provides northern Nigerian farmers with virtual farming advice (*Field 1*), *and* supplies them with agricultural inputs and markets their outputs (both *Field 2*), *and* offers them financial products (*Field 4*). Nor do multilateral organisations that support such ICT investments limit themselves to a single field. For example, in its strategy for the coming decade, FAO will work towards "accessible digital ICT technologies to enhance market opportunities, productivity and resilience integrated into agri-food systems policies and programmes, with particular focus on ensuring affordable and equitable access of poor and vulnerable rural communities".[22] This work is likely to cover at least *Fields 1, 2, 3, 6* and *7*. Reports that are eager to celebrate ICT's successes also (and often uncritically) present results across different fields. The 2021 *World development report*, for example, claims results in *Fields 1, 2* and *4* when it reports that

> in India, farmers can access a data-driven platform that uses satellite imagery, artificial intelligence [...] and machine learning [...] to detect crop health remotely and estimate yield ahead of the harvest. Farmers can then share such information with financial institutions to demonstrate their potential profitability, thereby increasing their chance of obtaining a loan.[23]

Notes

1 This is the summary paragraph of Bayer (January 2021) *Digital farming is driving sustainability*, Bayer Global. (Strikingly, for an article on environmental sustainability, the article's only picture is of a Brazilian farmer, his son, two grandsons, and his private plane.)
2 Prescott-Allen, R. *et al* (October 1991) *Caring for the earth; a strategy for sustainable living*, International Union for Conservation of Nature, United Nations Environment Programme (UNEP) and World Wildlife Fund (WWF, now World Wide Fund for Nature), with the quotation from Box 0.1 on page 10. UNEP has since shifted its position and nowadays its publications are more ecomodernist in nature.
3 McMichael, P. (2017) *Development and social change; a global perspective*, sixth edition, SAGE, with the quotation from page 10.
4 Fairphone is a Dutch social company that aims to develop smartphones that do not contain conflict minerals and that are designed and produced in a way that minimises their environmental footprint, also by using a modular construction that lengthens their productive life because it facilitates repair.
5 Huesemann, M. and Huesemann, J. (2011) *Techno-fix: why technology won't save us or the environment*, New Society Publishers, with the quotation from the second paragraph of the introduction.
6 Fukuyama, F. (1992) *The end of history and the last man*, Free Press.
7 World Bank (2008) *World development report; agriculture for development*, World Bank, page 247.
8 E.g., the worst assessment to date is presented in the 2023 synthesis report of the Intergovernmental Panel on Climate Change (IPCC): IPCC (2023) *Synthesis report of the IPCC sixth assessment report*, Intergovernmental Panel on Climate Change.
9 The world is enduring the highest number of conflicts since the creation of the UN, with one quarter of the global population now living in conflict-affected countries. United Nations (2022) *The sustainable development goals report 2022*, United Nations, pages 2–3. For some sobering statistics, see the background section of ICAI (December 2022) *The UK's approaches to peacebuilding*, International Commission for Aid Impact.
10 Such casual dismissals are common. This particular one comes, without substantiation or references, from a book review of Latouche's *Farewell to*

growth: Kallis, G. (March 2011) "The degrowth proposal", *Ecological Economics*, volume 70, issue 5, pages 1016–1017, with the quotation from page 1017.

11 The global financial system nearly collapsed in 2008, because too many banks had taken too much risk. This negatively affected global economic activity and trade, and compounded the effects of the food and fuel price hikes of the years preceding it. Together, the price hikes and financial crisis formed the 'triple-F crisis' of Food, Fuel, and Finance. For most adults, it led to extra but temporary hardship and stress. Some of this hardship continued after the global economy resumed its growth, as people had sold their productive assets to buy food or had incurred debts that were hard to repay. The effects were worst and most permanent for many children. A child that drops out of school does not automatically return to it in better times, and stunted growth cannot be reversed by eating more a few years later. One estimate, based on interviews with a random sample of 26,000 African mothers, concluded that the triple-F crisis had led to the deaths of an additional 28,000–50,000 infants in Sub-Saharan Africa alone. Most of them were girls. Friedman, J. and Schady, N. (2013) "How many infants likely died in Africa as a result of the 2008–2009 global financial crisis?", *Health Economics*, volume 22, pages 611–622.

12 This promise lumped migration together with "disease […], the threat of terrorism and global climate change" as the "great global challenges" that the UK aid strategy set out to tackle. HM Treasury and DFID (November 2015) *UK Aid: tackling global challenges in the national interest*, Her Majesty's Treasury and the Department for International Development, with the quotations from page 3.

13 For example, in 2015 the Dutch government earmarked €20 million of the Dutch Good Growth Fund for interventions in "migration countries", and several of these interventions had a strong ICT focus. See Itad (August 2020) *Evaluation of the Dutch Good Growth Fund; final report*, Itad, with the earmarking covered on page 33. On page 8 of this evaluation, I concluded that the Dutch government had misunderstood the causal link between people's prospects and their propensity to migrate: "The assumption that DGGF could reduce irregular migration flows to Europe by increasing opportunities to youth in a group of 'focus countries' was disproved by almost all empirical research. The opposite is true: in low- and lower middle-income countries, better prospects increase people's propensity to migrate."

14 Based on an unpublished Babban Gona presentation in March 2021. I attended this presentation in the context of a non-evaluative assignment, and I have not verified the information.

15 Itad (August 2020) *Evaluation of the Dutch Good Growth Fund; final report*, Itad, with the DGGF's three Theories of Change on pages 77–79. A Theory of Change is a planning and adaptive management tool that helps donors, NGOs and social enterprises monitor their progress after first defining what they intend to work towards (their long-term objectives), the in-between results needed to get there (the short-, mid- and long-term outcomes) and the broad strategy they hope to follow to make that happen.

16 I cannot provide the source because, at the time of writing this book, the statement is not yet in the public domain.

17 For its ever-changing list of partners, see pharmaccess.org/partners. ADB might be an error as it stands for and is linked to the website of the Asian Development Bank, but as far as I know PharmAccess works in Africa only. This is probably meant to be a reference to AfDB, the African Development Bank.

18 Bright, J. (18 January 2017) *Mastercard launches 2KUZE agtech platform in East Africa*, Extra Crunch. The Gates Foundation was never reluctant to engage with large transnational companies – it also partnered with Cargill and Syngenta (two giant agricultural companies, from the USA and Switzerland respectively), and bought Monsanto shares.

19 For example, Mann, L. (2017) "Left to other peoples' devices? A political economy perspective on the big data revolution in development", *Development and Change*, volume 49, issue 1, pages 3–36. Laura Mann does not mention 2KUZE but does cover some of Mastercard's work in Africa's e-commerce and agricultural sectors.

20 For example, Hannah McCarrick and Dorothea Kleine report on women-focused ICT training projects in rural Zanzibar and rural Chile that by and large failed to achieve the envisioned results of female entrepreneurship. But in both cases there was an arguably *somewhat* successful woman (the one in Chile very reluctantly so as she would have preferred to be in government employ), and each of them then "became a poster child for female entrepreneurship". See McCarrick, H. and Kleine, D. (2017) "Digital inclusion, female entrepreneurship, and the production of neoliberal subjects – views from Chile and Tanzania", chapter 4 of Graham, M., editor, *Digital economies at global margins*, The MIT Press, with the quotation from page 120.

21 Kranzberg, M. (July 1986) "Technology and history: 'Kranzberg's laws'", *Technology and Culture*, volume 27, number 3, pages 544–560, with the quotation from pages 545–546.

22 FAO (March 2021) *Strategic framework 2022–31*, Food and Agriculture Organization, with the quotation from Table 2 on page 16.

23 World Bank (2021) *World development report: data for better lives*, A World Bank Group Flagship Report, with the quotation from page 92.

1 Agricultural production processes

Millions of small-scale farmers are already using a range of recently introduced ICT-based tools that move them in the direction of 'digital agriculture'.[1] This range of tools is rapidly expanding, and so is the size of their user groups.

Some ICT tools merely facilitate 'traditional' contacts. Extension workers and vets can be in more frequent contact with farmers, and vice versa, if contacts are by phone than if they require face-to-face encounters. Apps such as FaceTime or WeChat also facilitate timely contact, and enable the farmers and extension workers to send each other photos and clips. Similarly, farmers have always learned from the people in their immediate social groups, and mobile phones facilitate communication within such groups.[2] Agricultural radio programmes have a long history in the Global South, and nowadays they often have phone-in and SMS options that make these programmes more interactive.[3]

In addition, social media expands people's networks far beyond what was traditionally possible.[4] Many farmers, hunters and foragers use general Facebook-type platforms for this, but there are also specialist farming channels. WeFarm, for example, dramatically increased farm-to-farm learning opportunities with its free SMS-based platform that connects farmers, originally in Kenya, Tanzania and Peru (and by now there are millions of them, or so WeFarm says).[5] Unlike channels where actual farmers directly communicate with one another, these specialist platforms use machine learning algorithms[6] on the basis of which farmers who ask questions are connected with other farmers, and receive answers via text messages or chat services. Because it is largely driven by farmers' input, WeFarm raises the visibility of indigenous knowledge, something often ignored by agribusiness and donors alike.

At least in Africa and at the time of writing, such text- and voice-based mobile phone services are, together with information

DOI: 10.4324/9781003451716-3

dissemination through radio broadcasting, still the main types of ICT applications.[7] However, this is gradually changing as 4G is expanding into rural regions, rural people are gaining ICT skill gains, and donors are funding and leveraging major investments in internet-based product development and marketing. Even WeFarm, which traditionally marketed itself as "the internet for farmers without the internet",[8] has moved onto an online presence after it received the necessary financing to launch an app, in March 2021.[9] This new app is merely one addition to a rapidly increasing number of agricultural apps,[10] often in appropriate languages[11] (which is key to utilisation[12]) that help online and sometimes largely offline farmers.[13] In relation to the agricultural production process, most of these apps do either one or both of two broad groups of things:[14]

- **Support farm management and precision farming.** Apps support – or at least *aim* and *claim* to support – farmers with a wide range of things. They advise on and facilitate crop choices (with a bias against economically less important crops – see below), input choices, planning, field preparation, access to financing, planting, weeding, feeding, watering, fertiliser management, pest control, harvesting, storage, and all other elements that are part of the agricultural cycle. In each of these fields, the range of support options is expanding. For water management, for example, farmers use localised weather apps, but increasingly also apps that are part of the 'Internet of Things' and linked to sensors that enable remote sensing of soil moisture and smart remote management and monitoring of the use of water.
- **Prevent and manage pests and diseases in plants and animals, as well as other risks.** For some fruits and vegetables, there are apps that use photos to identify pests and diseases immediately, and apps that invite farmers to send pictures of plants, which are then analysed for diseases, and alert farmers of threats – including collective threats, which presents opportunities for collective management of these threats. Apps warn people in case of immediate climatic and other dangers (e.g., locust swarms, volcano eruptions, a Boko Haram attack threat) and enable them to, for example, move with dependants and livestock to safe places before the storm reaches them. Apps provide micro-insurance products, which are then linked to satellite imagery and climate sensors to verify if pay-out criteria are met (because such criteria are not based on an individual's harvest but on, mostly, a region's overall precipitation and greenness – I return to this in Chapter 4).

In both fields, the functionality of ICT applications is advancing rapidly. Weather-related ICT, for example, started with regional radio and television reports, after which more localised text messaging facilities were launched in the late 20th century. This was followed by apps that track their users' GPS locations. These apps identify farmers' micro-locations and advise, in the morning, how much water their crops require that day, and if local pest risks might be on the rise and how they could be mitigated (all useful information that may, however, gradually cause indigenous systems and techniques to disappear – see below). The costs of agricultural ICT services are coming down as well. Remote irrigation management used to be for large farms only, for example, but Nano Air in Senegal now produces a low-cost remote irrigation management system (Widim Pump) that costs just €305.[15] In my book on rural development I wrote that, because of high costs, "small farmers in the Global South [...] don't get their cows chipped to monitor their health and fine-tune their feeding habits"[16] – but by the time the book was published, in June 2020, this was no longer the case as Jaguza had introduced affordable ear tags that monitor cows' health in Uganda.[17] These tags may also mitigate against the high risk of cattle theft in the north of the country, as they track the cows' whereabouts.

In addition to existing solutions advancing their functionality and reducing their costs, new ICT products and services are launched every month. Just before the lockdowns caused by the pandemic started, I visited a number of agritech start-ups in India – a major global player in ICT. Fasal and TartanSense seemed particularly promising. Fasal was about to launch in-field sensors that link up with farmers' smartphones to give real-time information about a particular field's crop and soil conditions. TartanSense was conducting final tests on robots that drive around small cotton farms with an eye and algorithms that are ever-evolving and based on big data, that enable these robots to spray weeds and diseased cotton plants, while leaving healthy plants undisturbed. As its next step, TartanSense was going to add precision-sowing functionality to its robots. A third Indian agritech start-up company that I visited in 2020, Yuktix, markets itself as a company that gives visibility to indigenous knowledge, helps verify what works and what does not, and helps indigenous knowledge to evolve and remain relevant in times of climate change and increasing water scarcity:

> Yuktix [...] helps in the verification of Indigenous Knowledge practices to bring out their true potential. The platform includes advanced sensors for micro weather measurement, soil moisture monitoring systems, and pests, and disease forecasting. That means

farmers and scientists have a tool to correlate indigenous knowledge with actual measurements. [...] The platform helps bridge the gap between researchers and farmers. While both have domain expertise, they seldom had a two-way audio-visual interaction channel for better interaction. Now farmers and scientists can easily share data and recommendations.[18]

ICT facilitates precision farming

These start-up companies may not survive. If they do, they may not deliver what they promise – a common problem I return to in Chapter 9. But if they do deliver on their promises, then the sensors of Fasal, the robots of TartanSense and the farmer–scientist interface of Yuktix will improve the efficiency of small farmers' farming in India and beyond, reducing their use of water and chemicals and increasing their harvests.

At country level, ICT-facilitated farming is not close to reaching that point. There are only a few country-level attempts to isolate the effects of ICT applications on agricultural productivity from the many other influencing factors,[19] and these studies suggest that, at country level, the ICT contribution to agricultural productivity is still modest. The most thorough piece of research focuses on China.[20] It finds that ICT contributed to agricultural productivity because of its enhancing effects on technical *efficiency* (i.e., it takes less time to look for, find and apply relevant information, throughout the agricultural cycle), but did not affect technical *progress* (i.e., ICT did not add to or spread the use of advanced technologies in the production process). The modesty of the overall effect might be illustrated by the fact that there is no kink in the long-term trend of China's upward-sloping crop production index graph. Between 1994 and 2018, agricultural production more than doubled while the amount of cultivated land in China barely increased, but it was steady, uninterrupted growth, rather than growth that accelerated when ICT reached the country's rural regions.[21] In an attempt to find interesting country case studies, I checked the crop production index graphs of a range of other countries, and I did not find a relatively sudden trend change anywhere.[22] This disappointed me, as most of the farmers I had met in recent years used at least some sort of phone or text messaging support – probably a consequence of my evaluation bias (explained in Box 1.1). I assume that the reason for the lack of clear ICT contribution at country level is that the *overall* level of rural ICT skills is still low – significantly lower than the average farmer I get to meet – and that this stands in the way of the adoption of advanced technologies.

Box 1.1 The common bias of evaluators and researchers

I sometimes come across documents of institutional donors and NGOs that argue that "farmers can gain productivity by modernizing the agricultural sector" but that this is not happening and "we seem to be stuck with traditional and outdated systems of work".[23]

Such statements do not align with my own experience, and I am inclined to dismiss them. It seems to me that we live in an era in which rural life in general and agricultural practice in particular are changing at unprecedented speed. However, my experience is unrepresentative of 'the average' agricultural practice, as evaluations are typically focused on pilot initiatives and other forms of innovation. Such statements do not align with the sense I get from academic research either. However, academic research probably has a bias that is comparable to my evaluation bias, as evolving farming practices have more research appeal than 'outdated systems of work'. As I wrote this book, I found it hard to correct for these 'confirmation errors'.[24] This is because I have no means to assess the extent to which they distort my or other researchers' sense of reality, and the extent to which we unknowingly look, in whatever we see, for confirmation that 'life is on the move'.

In all likelihood, rural life is changing everywhere, if only because of climate change and some near-universal ICT realities such as 2G and mobile telephony. However, many of the ICT-driven changes covered in this book currently impact rural life in *pockets* of the rural Global South only, rather than the rural Global South *in its entirety*. Nevertheless, even those pocket-specific changes are of value, as they give an indication of what is likely to come in ever-larger parts of the rural Global South.

The situation is more promising at regional and lower levels where particular areas or groups adopted ICT applications. Researchers who assessed groups for which there was reason to believe there might be ICT uptake, found that ICT applications have indeed often improved agricultural productivity, especially if they filled key information gaps in contexts with large information asymmetries.

Most such research looked at the impact of ICT-enabled extension services and text messages via mobile phones. Apoorv Gupta and his colleagues conducted particularly thorough research in this field, in India.[25] They compared areas that had gained recent phone coverage because a mobile phone tower had been requested and constructed, and areas where such a tower had been requested but had not been constructed. They then looked at survey data on agricultural inputs and crop yields, and they assessed the nature of the 2.5 million conversations that connected farmers had had, using a free agricultural advisory phone line (the governmental Kisan Call Centres). Especially in regions where the advisors and the farmers spoke the same language, they found that "areas where farmers had a larger increase in potential access to information experienced a larger increase in agricultural productivity".[26] They also found that the difference was verifiably the consequence of actions related to the topics discussed during these phone calls – often the appropriate use of pesticides and the choice of seeds. The paper lists other research on the effects of mobile phone-based extension services on farming practice and productivity, in India and elsewhere, most of which comes to comparable findings, though generally on the basis of research methods that were less thorough than those used by Gupta and his colleagues. Jenny Aker and her colleagues also reviewed literature and found that "results are mixed and, unsurprisingly, differentially distributed across the population" (as gender, caste and ethnicity matter). However, their overall conclusion was that "there is a growing body of economic evidence on the impacts of ICTs on agricultural information, knowledge, adoption, and prices [...]. The sum total of this evidence generally indicates that mobile phones in general, and ICT for agriculture more specifically, are having positive effects on targeted participants in specific agricultural markets".[27]

Notes

1 Also referred to as 'smart farming' or 'precision farming', which is hoped to lead to an 'agriculture 4.0' in which an amalgamation of technologies aims to help resolve whatever agricultural and climatological challenges might appear.
2 The prevailing view is that ICT greatly expands networks, but in some cases the most significant effect is the facilitation of contacts within pre-existing networks. When Petr Matous and his colleagues distributed mobile phones to a few hundred farmers in Ethiopia and tracked their use, they found that the farmers rarely used these phones to contact people they did not know. Instead, they used their phones to contact people who were in their social networks already, or who had been introduced to them as part of this research. Matous, P., Todo, Y. and Ishikawa, T. (2014) "Emergence of multiplex mobile phone communication networks across rural areas: an Ethiopian experiment", *Network Science*, volume 2, issue 2, pages 162–188.
3 E.g., Hudson, H.E. *et al* (2017) "Using radio and interactive ICTs to improve food security among smallholder farmers in Sub-Saharan Africa", *Telecommunication Policy*, volume 41, pages 670–684.
4 For an overview of ICT channels that could help preserve and disseminate indigenous knowledge (none of which are unexpected), see Owiny, S.A., Mehta, K. and Maretzki, A.N. (2014) "The use of social media technologies to create, preserve, and disseminate indigenous knowledge and skills to communities in East Africa", *International Journal of Communication*, volume 8, pages 234–247.
5 WeFarm started operations in 2015 but its reach and impact have, to the best of my knowledge and according to Godfrey Omulo and Eric Mensah Kumeh, not yet been independently assessed. See Omulo, G. and Kumeh, E. M. (2020) "Farmer-to-farmer digital network as a strategy to strengthen agricultural performance in Kenya: a research note on 'Wefarm' platform", *Technological Forecasting & Social Change*, volume 158, 6 pages.

6 An algorithm is a set of step-by-step procedures, or a set of rules to follow, for completing a specific task or solving a particular problem.

7 Ayim, C. *et al* (2022) "Adoption of ICT innovations in the agriculture sector in Africa: a review of the literature", *Agriculture & Food Security*, volume 11, issue 22.

8 Quoted in, among other places, Munford, M. (19 August 2013) *20 inspiring start-ups aiming to improve the world*, Mashable.

9 For WeFarm in general, see wefarm.com. For the new funding that allows for WeFarm's expansion into an online platform, see WeFarm (9 March 2021) *News release: $11 million to expand into online; WeFarm expands platform connecting the world's small-scale farmers*, WeFarm.

10 Julierme Zimmer Barbosa and his colleagues found 244 agricultural apps in 2015, and 599 in 2018, globally. Barbosa, J.Z. *et al* (2020) "Global trends for apps in agriculture", *Multi-Science Journal*, volume 3, issue 1, pages 16–20. This figure is a substantial underestimation of what exists in reality, because the research is based on a search in the Google Play and Windows Phone app stores, and the researchers seemed unaware that the offer of these app stores depends on the country you are conducting the search in. For example, on page 18 the paper concludes that there are more agricultural apps in Brazil than in the rest of the Global South combined, but they came to this finding simply because three of the four authors were conducting their research in Brazil.

11 To give but two examples: iCow, a mobile phone platform (and since 2018 a smartphone app) that provides livestock and crop advice and links farmers with a range of other players in the agricultural ecosystem, is available in English and Kiswahili in Kenya and Tanzania, and in Oromiffo, Amharic and Tigringnia in Ethiopia (see icow.co.ke/about); and Sunidhi Sharma and her colleagues found India-focused apps in Bengali, English, Gujarati, Hindi, Kannada, Malayalam, Marathi, Odiya, Punjabi, Tamil, Telugu and, in the case of an app called Rml kisankrishi mitr, "all the regional languages" (possibly referring to the 22 official languages that are recognised in the Constitution of India). Sharma, S., Sharma, D.K. and Sharma, S. (August 2018) "Overview of mobile Android agriculture applications", *International Research Journal of Engineering and Technology*, volume 5, issue 8, pages 225–231, with the overview of languages, per app, on pages 226–227.

12 For the importance of language, see Gupta, A., Ponticelli, J. and Tesei, A. (May 2020) "Information, technology adoption and productivity: the role of mobile phones in agriculture", *NBER Working Paper 27192*, National Bureau of Economic Research. This paper reports on a large-scale research effort focused on the Kisan Call Centres (KCC) in India, and the relative effectiveness of KCC's agricultural advisory services in regions where KCC provides services in people's first languages versus where it does not, and finds the differences in effectiveness are significant. The KCC is not app- but phone-based, but the principle is the same. There are many other papers that emphasise the importance of local languages (e.g., Ireri, D.M. (April 2020) "Influence of ICT weather forecasting on agricultural productivity in Kenya: a literature based review", *Journal of Information and Technology*, volume 4, issue 1, pages 56–69).

13 For example, GeoFarmer, an app that facilitates knowledge exchange among farmers (among many other things), is one of many apps that is designed to work offline and then updates whenever this is possible.

14 This is just my own and very basic categorisation. In addition to market apps (which I cover in Chapter 2), Sunidhi Sharma and her colleagues distinguish information-based apps, weather apps, advisory apps and management apps; Hetal Patel and Dharmendra Patel distinguish business apps, conference apps, apps for diseases and pests, farm management apps, learning and reference apps, location-based apps, and weather apps. See, respectively, Sharma, S., Sharma, D.K. and Sharma, S. (August 2018) "Overview of mobile Android agriculture applications", *International Research Journal of Engineering and Technology*, volume 5, issue 8, pages 225–231, with the list on page 228; and Patel, H. and Patel, D. (March 2016) "Survey of Android apps for agriculture sector", *International Journal of Information Sciences and Techniques*, volume 6, number 1–2, pages 61–67, with the list on pages 63–64. Other authors distinguish even more categories.

15 Ngounou, B. (9 May 2019) *Senegal: startup puts technology at the service of agro-ecology*, Afrik21.

16 Eekelen, W. van (2020) *Rural development in practice; evolving challenges and opportunities*, Routledge, with the quotation from page 160.

17 See jaguzafarm.com.

18 The quotation is from a Yuktix mailshot: Yuktix (22 June 2021) *A method to utilize indigenous knowledge in agricultural practices*, Yuktix. Yuktix elaborates and outlines examples of ways in which ICT in general and Yuktix in particular could support the utilisation and evolving nature of indigenous knowledge in Jha, R. (2021) *Agricultural indigenous knowledge (AIK) systems*, Yuktix Technologies Agriintelligence Solutions.

19 For example, Qiubo, Z. *et al* (November-December 2020) "Do ICTs boost agricultural productivity?", *China Economist*, volume 15, number 6, pages 9–26; Chatterjee, A. (2020) "Role of ICT in sustainable agricultural development – the case of India", chapter 16 of Sikdar, S., Das, R.C. and Bhattacharyya, R., editors, *Role of IT-ITES in economic development of Asia*, Springer.

20 Qiubo, Z. *et al* (November–December 2020) "Do ICTs boost agricultural productivity?", *China Economist*, volume 15, number 6, pages 9–26.

21 See data.worldbank.org/indicator, indicators of Crop production index and China.

22 See data.worldbank.org/indicator, indicators of *Crop production index* and individual countries. Note that the database has an aggregation problem and continental graphs show a very high decline in 2010–12 that did not occur in reality.

23 This particular quotation is from the introduction of an unpublished draft strategy document of a European donor, which outlines its contribution to sustainable economic development.

24 See chapter 5 ("Confirmation shmonfirmation!") in Taleb, N.N. (2008) *The Black Swan; the impact of the highly improbable*, Penguin. (The chapter title does not contain a spelling error.)

25 Gupta, A., Ponticelli, J. and Tesei, A. (May 2020) "Information, technology adoption and productivity: the role of mobile phones in agriculture", *NBER Working Paper 27192*, National Bureau of Economic Research.

26 *Ibid*, with the quotation from page 3.

27 Aker, J.C., Ghosh, I. and Burrell, J. (2016) "The promise (and pitfalls) of ICT for agriculture initiatives", *Agricultural Economics*, volume 47, supplement, pages 35–48, with the quotation from page 40.

2 Trade

When I studied economics in the 1980s, neo-classical economic theory was still part of the curriculum. I remember ridiculing this theory as it assumes that trade is based on equal and perfect access to information, which was obviously and painfully untrue. *Free* trade was rarely *fair* trade, if only because of the massive difference in the information small producers and large buyers had at their disposal. ICT makes neo-classical economics look potentially relevant again, at least at micro-economic level, and farmers often consider "market information, including daily updates on the prices of agricultural commodities in the local markets of the surrounding district, [to be] one of the most relevant ICT services".[1] Two key potential benefits of improved market information are the marketability of crops, and the price fetched for them.

Marketability. Saher Asad researched the effects of ICT on crop choices by comparing those of Pakistani farmers within 10 kilometres of the Indian border, where mobile phone coverage is forbidden, with farmers living right next to them, but outside of the exclusion zone.[2] He found that the farmers with access to mobile phones were more inclined to grow very perishable – but also more valuable – crops than their phoneless neighbours. His paper starts with this instantly clarifying quotation from a farmer: "Before I had a cell phone I harvested my crops and then had to wait for a trader to buy my crops; now I talk to the trader and harvest my crops when he will buy them". In a similar vein, banana-producing farmers in remote rural regions in Uganda increased their market participation after their region had been connected to mobile networks, because the real-time market information reduced the risks of attempting to sell perishable fruits to faraway markets.[3]

Price. Many farmers, fisherfolk and pastoralists no longer rely on a single local market or a single broker. Instead, they compare demand and prices across different markets, phone a few other middlemen, and check the prices offered by nearby factories. They may also use a text messaging

DOI: 10.4324/9781003451716-4

service that provides market price information. More recently, market app appeared, which offer customised and more precise versions of that same information. They may have a search function based on the user's location, for example, or engagement features that encourage price reporting so as to triangulate data across multiple sources, or bidding and match-making features. Examples include Esoko (formerly TradeNet) in a range of African countries,[4] and RML (or 'RML Information Services', which used to be called Reuters Market Light), Agri market and Iffco Kisan in India, where such apps seem to be more diverse and mature than in any other part of the Global South.

For products as wide-ranging as grain and sardines, studies find that price variation (*across markets*, not necessarily *over time*) have reduced with the introduction of mobile phones.[5] Robert Jensen famously illustrated this effect in a set of graphs that show that the early morning beach price of sardines in north Kerala fluctuated wildly before the introduction of mobile phones, and barely at all after it.[6]

The reason for such a price stabilization effect is simple: if the price offered by a buyer is too low, and farmers and fisherfolk are not obligated to sell to that buyer, then they will not do it, as ICT provides them with alternative outlets. This principle even applies after people have already arrived at a specific market:

> Mobile phones are [...] an essential tool for doing business at the weekly market. [...] both buyers and sellers use mobile phones to monitor changes in prices throughout the region, especially during the great sheep feast (Ayd Al Kabir), the most important family and religious feast in the country, where more than 4 million animals are sold in Morocco, helping both to avoid unreasonable prices.[7]

Such services are not yet readily and knowingly available everywhere. Research in rural Ethiopia found that there was little difference between prices fetched by farmers with and farmers without mobile phones, in six out of the seven crops the researchers considered. This was largely because farmers (and especially older farmers) did not know how to use their phones to access relevant market information.[8] Other studies found that available information about prices cannot always be *used*, even if it is *available*. Jenna Burrell and Elisa Oreglia found that Ugandan fishermen, or at least the ones in the village where they conducted their research, often sold their fish to the middlemen from whom they had received credit, without using their phone to check for other possibilities, because of the long-term importance of the relationships.[9] Even in Kerala, the region where Jensen conducted his sardines-focused research, later studies showed that not all fishermen are free to sell their fish to the highest bidder,[10] and that the results of Jensen's research are therefore not as generalizable as the World Bank and other lobbyists for ICT4D present them to be.[11] Similarly, potato producers in West Bengal could not translate their improved knowledge of prices into better prices for themselves, because wholesalers did not want to buy directly from farmers and only ever dealt with middlemen;[12] and in the central Himalayas, too, farmers' "exploitation is less about scarcity of information and more about scarcity of choice".[13]

Where production and market information *are* available and buyers and sellers *are* at liberty to choose each other, the interplay between buyers and sellers, and the usage of data, can be very complex indeed, and small-scale farmers, fisherfolk and pastoralists are not necessarily the ones reaping the benefits. They are able to check market prices, but so are the buyers. Moreover, buyers may have the advantage of additional ICT-powered information that helps them predict the likely size of the coming harvest (through satellite imagery) or the size and nature of the fisherfolk's catch (through shipping tracking technologies). The most extreme case of the data-informed interplay between buyers and sellers comes from the Icelandic fishing industry – admittedly rather far from the Global South. Here, automatic and continuous tracking is obligatory for all ships. Using an open-access website (*marinetraffic.com*, the 'Facebook of maritime affairs'), buyers monitor all ships simultaneously, and they place their bids in virtual auctions[14] on the basis of, among other criteria, a ship's location (for instance, buying haddock but not cod from a ship that has fished inside fjord X because the haddock there is good but the cod from there is infested with parasites). Fisherfolk in turn anticipate these buyers' preferences, *and* monitor the whereabouts of other ships ("you can see where all the boats are" and "everybody [therefore] knows [a fisherman's] secret spots"), and plan their next trips accordingly.[15]

Other apps use yet other methods that may benefit trading efforts. In China in particular, e-commerce is large (amounting to over US$1.5 trillion in 2020, of which a sixth took place in rural regions)[16] and booming. It includes a rapidly growing trade in agricultural products, and this growth accelerated further during the Covid-19 pandemic.[17] Agricultural e-commerce includes warehouse-based e-commerce, but supply and demand also find each other directly ('farm-to-kitchen') via WeChat, Pinduoduo and Alibaba, where buyers are clients such as restaurants and individual households, and sellers are individual farms, farm groups and cooperatives. Increasingly, apps serve as a crowdsourcing marketplace, where apps create 'one big virtual farm'[18] by bulking the produce of farmers and farm groups, and the collective need for agricultural inputs, to negotiate better prices (e.g., MahaFarm, Babban Gona, Khula). Their growth is underpinned by a range of quality assurance measures that give online customers confidence. This is sometimes supported by government:

> In Shandong Province, the Shouguang Municipal Bureau of Agriculture created a quality tracing system for peppers and other vegetables, which farmers can avail of for free. Using a mobile app to scan the [...] QR code, customers can access information such as cultivation base; sampling time; results of pesticide concentration test planting (pruning, splitting, and watering); harvest; and sales transaction data. Access to such information helps strengthen customers' confidence in the product's quality and brand.[19]

Other ICT-powered quality assurance measures are related to the production process. In China, for example, Taobao Live instils consumer confidence by enabling farmers to livestream the entire agricultural cycle. And in Senegal, Pix Fruit assesses a farmer's likely delivery capacity by estimating the farmer's mango harvest size. Pix Fruit does this by combining a farmer's photos of a sample of trees with fruit-recognition technology and big data that the app's developers have collected with drones. These more reliable harvest size estimates help the farmer to bargain for a better price.[20]

Such market-facilitating ICT applications fit within the overall free trade paradigm, and therefore appeal to Western donors and some governments in the Global South. These donors and governments support the development and roll-out of such applications, and praise them in their publications as 'inclusive market development'. For obvious commercial reasons, agritech companies are similarly positive about the achievements of their products and services. The website of the above-mentioned RML, for example, has an 'impact' section that lists significant achievements, with reference to five studies, "conducted by various independent organizations" – but all the studies are at least a decade old, and the 'independent organizations' include the not-quite-independent World Bank and USAID.[21]

Less biased publications tend to be less positive. The only rigorous and independent piece of RML-focused research I am aware of did not come to positive conclusions (and is not included in the list on RML's website). It used a randomized controlled trial in a hundred Indian villages and concluded that:

> We find no statistically significant average treatment effect on the price received by farmers. [...] Treated farmers appear to make use

of the RML service and they associate RML information with a number of decisions they have made. But, based on the available evidence, on average they would have obtained a similar price or revenue, with or without RML.[22]

While the conclusions of research on price- and market-related agricultural ICT applications are less positive than the World Bank and the various company and donor websites might suggest, they tend to be more positive than the study just quoted. They range from neutral to significantly positive, and the more positive of the literature reviews come to conclusions such as this one:

> The research reviewed suggests that new information technologies involving mobile communications and networks are realising significant benefits of speed, mobility and efficiency in information exchange, creating new opportunities for arbitrage and making it easier for small producers and new entrants to participate and compete in markets.[23]

This result is probably a lot better than the previous generation of donor investments in 'M4P' programming (sometimes called 'MMW4P' programming – both stand for Making Markets Work for the Poor). M4P programmes I assessed rarely had lasting effects. Their claims ("so far the project has supported 332,634 people") were unverifiable but unlikely.[24] One of the more credible 'success stories' required a programme spend that was five times the claimed income gains over a five-year period (using a zero Net Present Value discount rate and assuming that the entire income differential was the consequence of the M4P intervention). Other M4P programmes (including ones that I did not assess myself) did not seem to have many achievements to report on either, as they frequently regurgitated the same few success stories. I recall an M4P's quarterly newsletter in Bosnia and Herzegovina celebrating the same tomato farmer in the same manner in three consecutive newsletters, for example, and another programme's reporting regularly revisited the story of a successful tangerine farmer in East Africa. There may be two reasons for the M4P efforts' lack of success. First, a lot of the M4P programming was naïve. It was usually designed by generalist NGO staff, rather than by market specialists, and its attempts to create or enter markets often proved to be either unfeasible (because of the lack of demand or the trading obstacles and costs) or unreliable (because the demand was too erratic for small-scale farmers). I saw NGOs piloting crab farming without crab markets, and supporting the organic production of unusual vegetables for a fragile niche within the hospitality industry. Both were NGO-implemented M4P programmes, and the demand for both products proved to be unreliable. I myself have been guilty of such naïve M4P programming as well – such as when I arranged a training programme for rural women in north Pakistan to make hand-drawn '*Happy Eid!*' picture cards, and set up a distribution system for them in the UK, only to find it impossible to sell these cards after a group of pity buyers had bought a first batch. Second, it may be easier to tailor small-scale farmers' behaviour to align with the needs of corporate value chains than it is to tailor markets to small-scale farmers. As Sally Brooks writes, "the intractable problem of 'making markets work for the poor' [gave] way to the more straightforward one of making more effective market subjects".[25]

All in all, and on average, donor investments in market-facilitating ICT applications probably represent better value for money than the average pre-ICT M4P programme has done. However, benefits differ widely across farmers. Much depends on their profile – and especially on their primary language and age, their ICT and general literacy levels, their region's connectedness and their purchasing power. Much also depends on the nature of the crops they grow. The main advantage for perishable crops appears to be that these crops become more easily marketable (a particular challenge during the pandemic – see Box 2.1). The main advantage for less perishable crops is probably that these applications increase the farmers' bargaining power[26] – provided that farmers feel at liberty to bypass their traditional buyers. In both cases, ICT has been most useful for farmers who grow crops that are of high economic value (from the point of view of the larger agro-industry), as these are the crops that ICT product development has focused on.[27] For a farm to benefit from such products, it may have to be incorporated into a global value chain. Such incorporation is heavily pushed by donors, multilateral agencies and the transnational agro-business alike. Critics say this concentrates the power into the hands of a small number of global companies[28] – an issue I will return to in Chapter 9.

Box 2.1 Market apps during the Covid-19 pandemic

The pandemic changed markets for agricultural produce around the world as sales to offices, schools and the hospitality industry plummeted, transport was restricted and borders were closed. The impact was often dramatic. In the last week of February 2020, I was on the Ethiopian side of Lake Turkana, where I saw villagers drying fish and selling it to traders who transported daily truckloads full of dried fish to markets in neighbouring countries. A mere month later, the pandemic had caused the Ethiopian government to close the country's borders and my interpreter got in touch to lament the results of that closure. He told me that most villagers had lost their main source of income, and that many were already facing food shortages as their savings had run out and the local food markets – which were normally stocked by the trucks that brought fresh produce on their way back – were now empty.

I have not yet seen any research papers about the ways in which market apps facilitated trade during the pandemic,[29] but I expect their role to have been significant. They could not help the Ethiopian fish-only village economies that were entirely dependent on cross-border trade, but farmers and fisherfolk in more diverse economies may have used the real time data of market apps to find markets that remained open. At least one app – Kisan Rath from India – was introduced *because* of the pandemic, to help maintain long-distance value chains in the midst of transport restrictions, by enabling farmers to book space in trucks that were still allowed to travel.

Notes

1 This particular quotation presents a common conclusion. It comes from Dhaka, B.L. and Chayal, K. (September 2010) "Farmers' experience with ICTs on transfer of technology in changing agri-rural environment", *Indian Research Journal of Extension Education*, volume 10, number 3, pages 114–118, with the quotation on page 116.

2 Asad, S. (March 2016) *The crop connection: impact of cell phone access on crop choice in rural Pakistan.* (The paper does not provide further reference information.)

3 Muto, M. and Yamano, T. (December 2009) "The impact of mobile phone coverage expansion on market participation: panel data evidence from Uganda" *World Development*, volume 37, issue 12, pages 1887–1896.

4 In May 2019 the countries in which Esoko was operational were Benin, Burkina Faso, Ghana, Kenya, Madagascar, Malawi, Mozambique, Nigeria and Zimbabwe. Esnault, D. (5 May 2019) *Smart tech the new tool for African farmers*, Phys.Org.

5 These and a few other examples are reported on in Duncombe, R. (2016) "Mobile phones for agricultural and rural development: a literature review and suggestions for future research", *European Journal of Development Research*, volume 28, pages 213–235.

6 Jensen, R. (August 2007) "The digital provide: information (technology), market performance, and welfare in the South Indian fisheries sector", *The Quarterly Journal of Economics*, volume 122, issue 3, pages 879–924.

7 Vidal-Gonzáles, P. and Nahhass, B. (2018) "The use of mobile phones as a survival strategy amongst nomadic populations in the Oriental region (Morocco)", *Geojournal*, volume 83, pages 1079–1090, with the quotation on page 1086. The text mentions that researchers in Mauritania had made the same observations.

8 Tadesse, G. and Bahiigwa, G. (2014) "Mobile phones and farmers' marketing decisions in Ethiopia", *World Development*, volume 68, pages 296–307. Farmers with mobile phones did fetch higher prices for wheat (the seventh crop).

9 Burrell, J. and Oreglia, E. (2015) "The myth of market price information: mobile phones and the application of economic knowledge in ICTD", *Economy and Society*, volume 44, issue 2, pages 271–292, with the issue covered on pages 281–282. A shorter version of this paper is included, under the same title, as chapter 7 of Graham, M., editor (2017) *Digital economies at global margins*, The MIT Press.

10 Srinivasan, J. and Burrell, J. (2015) "On the importance of price information to fishers and to economists: revisiting mobile phone use among fishers in Kerala", *Information Technologies & International Development*, volume 11, issue 1, pages 57–70.

11 World Bank documents have frequently quoted and referenced Jensen's 2007 paper, and its 2016 World development report reprinted his graphs. World Bank (2016) *World development report: digital dividends*, A World Bank Group Flagship Report, Figure 1.1 on page 91.

12 Mitra, S. *et al* (March 2018) "Asymmetric information and middleman margins: an experiment with Indian potato farmers", *The Review of Economics and Statistics*, volume C, issue 1, 13 pages.

13 Payal Arora, when explaining why "the 'information highway' is more a side street in the process of decision-making"; see page 94 of Arora, P. (2016) *Dot com mantra: social computing in the central Himalayas*, Routledge (first published in 2010, by Ashgate Publishing).

14 Virtual auctions are on the rise, also in the Global South. There are online tea auctions in India and Sri Lanka, for example. However, such virtual trade requires a level of trust and quality assurance, as buyers will not otherwise buy produce without physical inspection or recommendation from trusted intermediaries. This means that, compared to physical auctions, online auctions might be biased towards the larger suppliers, as such trust and quality assurance measures are easier to build with a few large suppliers than with many small-scale ones. Moreover, such online auctions are hotly resisted by intermediaries who see their interests threatened. For a case study of such resistance, ultimately leading to the failure of a virtual tea auction experiment in Kenya, see Foster, C., Graham, M. and Waema, T.M. (2017) "Making sense of digital disintermediation and development: the case of the Mombasa Tea Auction", chapter 2 in Graham, M., editor, *Digital economies at global margins*, The MIT Press. The chapter also elaborates on the other points made in this endnote.

15 Dobeson, A. (2016) "Scopic valuations: how digital tracking technologies shape economic value", *Economy and Society*, volume 45, issue 3–4, pages 454–478, with the quotations from page 463.

16 MOFCOM (24 January 2021) MOFCOM *Department of Electronic Commerce and Informatization on the development of the online retail market in*

2020, Ministry of Commerce of the People's Republic of China. The figures reported are in yuan (or rather in 'renminbi', or RMB, the yuan's official equivalent) and are RMB 11.76 trillion and RMB 1.79 trillion respectively.

17 Moore, L. (25 November 2020) *Rural ecommerce: in China, farmers find new ways to grow*, SupChina.

18 This phrase of 'one big virtual farm' is from Khula. See, for example, Gilbert, P. (12 October 2018) *Agricultural app Khula wins MTN Business App of the Year*, ITWeb telecoms.

19 ADB (September 2018) *Internet Plus agriculture; a new engine for rural economic growth in the People's Republic of China*, Asian Development Bank, with the quotation from page x and a more elaborate text on page 24.

20 Esnault, D. (5 May 2019) *Smart tech the new tool for African farmers*, Phys. Org.

21 rmlagtech.com; go to *About us*, then *Impact*, accessed on 20 May 2021; no longer available in September 2023.

22 Fafchamps, M. and Minten, B. (2012) "Impact of SMS-based agricultural information on Indian farmers", *The World Bank Economic Review*, volume 26, number 3, pages 383–414, with the quotation from page 412.

23 Duncombe, R. (2016) "Mobile phones for agricultural and rural development: a literature review and suggestions for future research", *European Journal of Development Research*, volume 28, pages 213–235, with the quotation on pages 230–231.

24 The quotation is from a performance report of an M4P programme that was implemented in East Africa. I assessed this programme and was unable to trace this number back to actual people, or to understand what method had been used to arrive at this number.

25 Brooks, S. (2021) "Configuring the digital farmer: a nudge world in the making?", *Economy and Society*, volume 50, issue 3, pages 374–396, with the quotation on page 375. (Brooks makes this observation bitterly, as she takes issue with the incorporation of small-scale farmers into large value chains.)

26 Sometimes this shift in bargaining power even spills over to farmers who do not themselves have access to price information. This was the case for yams in Ghana and is described in Hildebrandt, N. *et al* (September 2020, revised in February 2021) *Price information, inter-village networks, and "bargaining spillovers": experimental evidence from Ghana*, NYU Stern School of Business.

27 Visser, O., Sippel, S.R. and Thiemann, L. (August 2021) "Imprecision farming? Examining the (in)accuracy and risks of digital agriculture", *Journal of Rural Studies*, volume 86, pages 623–632.

28 Mann, L. (2017) "Left to other peoples' devices? A political economy perspective on the big data revolution in development", *Development and Change*, volume 49, issue 1, pages 3–36, with the most relevant bits in pages 18–21.

29 There are a few publications with titles that suggest they cover this issue, but instead of reporting research findings they all merely *predict* that market apps will prove to be particularly useful during the pandemic. See Mahapatra, S.K. (2020) "Smartphone apps for agri-information dissemination during Covid19 lockdown", *Biotica Research Today*, volume 2, number 5, pages 116–119; and Dewi, D.A. and Abdullah, Z.H. (November-December 2020) "A review on agricultural mobile apps for sustainable agribusiness: before and during Covid-19 pandemic", *International Journal of Innovation Engineering and Science Research*, volume 4, issue 6, pages 53–57.

3 Diversified livelihoods

Some ICT products and services have caused a rise in rural incomes of households that use them. In part, this is because ICT products and services sometimes enable farmers to improve their production and to gain easier access to better-paying markets. In addition, some ICT products and services increase non-farm income earning opportunities, because they match the demand for and supply of labour and services. Rebecca Hartje and Michael Hübler showed this for mobile phone and smartphone users in the Mekong region of Thailand, Vietnam, Laos and Cambodia.[1] Wanglin Ma and his colleagues showed it for smartphone users and internet users in rural China.[2] I am not aware of any such research having been conducted anywhere in Africa[3] but I think the phenomenon exists there as well, as I've come across many rural people who gain some of their income by offering their spare room, their car or their interpreting services through virtual channels.

In addition to income gains, ICT may help rural households to reduce risks by *diversifying* their sources of income. A few of the ways in which this could happen are covered in the previous sub-sections: as ICT enables farmers to increase their production and improve their marketing, they may try new crops, increase the portion of their farm that is dedicated to cash crops, and diversify their outlets. Another part of the explanation is that ICT – and mobile money in particular – facilitates safe participation in the monetised part of the economy. ICT solutions also reduce the transaction costs incurred when buying or selling goods and services, and ICT expands the market for nearly every product or service people may have on offer. This is because ICT makes it easier for demand and supply to find each other, either with the help of a specialist app such as Kisan Rath or Hello Tractor, or simply by making a few mobile phone calls. Consequently, spare capacity (storage space, transport capacity, extra cooking, a tractor), specialist skills (to fix bicycles, to bake cakes, to DJ) and spare time (to clean a house, to work on another farmer's field, to participate in a public works programme) are all more easily utilised. This

DOI: 10.4324/9781003451716-5

builds onto most poor rural people's long-standing strategy of income diversification (see Box 3.1).

Box 3.1 The poor rural household's multiple sources of income

When, in the course of livelihoods project evaluations, I asked rural people about their source of income, the initial answer was generally straightforward.

I farm / I am a farm worker / I repair bicycles / I am a teacher

Early on in these evaluations, I learned it was useful to then ask this:

Oh, ok. Thanks. What else?

If I was then quiet, patiently waited for the next answer and then repeated the question a few more times, I almost invariably found that the people I was listening to spread their bets.

Sometimes my uncle sends us money. I also work on the cotton farm down the road. Sometimes I go fishing. I get fruits and mushrooms from the forest. I collect firewood, and burn and sell charcoal. I hunt for bush rats. I dig for gold sand and quarry stones.

Rural households are rarely single-purpose users of the commons around them.[4] Instead, they use a range of common goods – and this use is higher when they are poor or when times are hard. At the end of the agricultural cycle or when an income disappears because of illness or a death in a family, the forest and river provide a safety net of sorts, and people turn to them for marketable products and consumption.[5]

Such diversification is helpful for a range of reasons.

Diversification of income sources is beneficial to rural households in developing countries because it helps mitigate household income variability, address external shocks in production (e.g., unpredictable rainfall and drought, crop diseases, or a steep decline in farm produce prices), smoothen household consumption patterns, and reduce livelihood vulnerability. [...] Rural household income diversification exerts a positive effect on on-farm liquidity constraints, which enabled rural households to pursue more lucrative livelihood strategies and improve their living standards.[6]

A further explanation for the effects of ICT on rural income diversification is that ICT facilitates savings, which can then be invested in new or better income-generating activities, as well as access to microfinance, which potentially facilitates investments through credit and reduces risks through insurance, among other things. The next chapter looks at ICT-facilitated savings and microfinance.

The nature of and reasons for the correlation differ across contexts, but the fact that there *is* correlation between ICT and income diversification has been confirmed in research:

- in China, where research identified a well-evidenced causal link in rural regions (and the opposite effect for towns and cities) and found stronger effects for poorer farmers than for wealthier ones;[7]
- in Rwanda, where the causal claim is less well-founded and where the diversification effects were found in rural and urban areas alike;[8]
- in India, where research confirmed correlation but not necessarily causality;[9]
- in Tanzania, where research also confirmed correlation but not necessarily causality;[10] and
- in the Mekong region, where mobile phones helped subsistence farmers increase their participation in the local off-farm labour market – and the benefits were such that this reduced their propensity to migrate to the city.[11]

ICT facilitates the diversification of livelihoods

Notes

1 Hartje, R. and Hübler, M. (March 2015) "Are smartphones smart for economic development?", *Hannover Economic Papers*, number 555.

2 Ma, W., Grafton, R.Q. and Renwick, A. (2020) "Smartphone use and income growth in rural China: empirical results and policy implications", *Electronic Commerce Research*, volume 20, pages 713–736; and Ma, W. *et al* (2020) "Impact of Internet use on economic well-being of rural households: evidence from China", *Review of Development Economics*, volume 24, pages 503–523.

3 There are publications that argue that incomes must have increased by pointing at research that addresses related issues (e.g., market efficiency, price stability) – such as Aker, J.C. and Mbiti, I.M. (2010) "Mobile phone and economic development in Africa", *Journal of Economic Perspectives*, volume 24, issue 3, pages 207–232. However, I do not know of research that *directly* looked at ICT's effects on income size in any of Africa's rural regions.

4 Zulu and Richardson looked at charcoal production in Sub-Saharan Africa, for example, and found that, for the vast majority of the people who make an income from the production and sale of charcoal, this is not their sole or even their primary source of income and that, instead, it "helps bridge seasonal gaps in income for farmers and helps generate working capital after clearing land in preparation for planting at the start of a new agricultural year. Charcoal also provides a 'safety net' in times of hardship". Zulu, L.C. and Richardson, R.B. (2013) "Charcoal, livelihoods, and poverty reduction: evidence from sub-Saharan Africa", *Energy for Sustainable Development*, volume 17, pages 127–137, with the quotation on page 132.

5 Falconer (1990) "'Hungry season' food from the forests", *Unasylva*, volume 141, number 160, pages 14–19.

6 Many publications offer similar overviews. This particular quotation is from Leng, C. *et al* (2020) "ICT adoption and income diversification among rural households in China", *Applied Economics*, volume 52, number 33, pages 3614–3628, with the quotation from page 3614.

7 *Ibid*. This piece of research is more thorough than any of the other publications covering this field.

8 Aristide Maniriho and Pia Nilsson used two consecutive 'Integrated Households Living Conditions Surveys' to consider a range of factors that could conceivably affect livelihood diversification, and concluded that access to ICT was one of the three most significant factors for income diversification (the other two were levels of education and residing in an urban setting). Maniriho, A. and Nilsson, P. (2018) "Determinants of livelihood diversification among Rwandan households: the role of education, ICT and urbanization", *East Africa Research Papers in Economics and Finance*, number 2018–24.

9 Khan, W., Tabassum, S. and Ansari, S.A. (2017) "Can diversification of livelihood sources increase income of farm households? A case study in Uttar Pradesh", *Agricultural Economics Research Review*, volume 30, pages 27–34.

10 Baird, T.D. and Hartter, J. (2017) "Livelihood diversification, mobile phones and information diversity in Tanzania", *Land Use Policy*, volume 67, pages 460–471.

11 Hartje, R. and Hübler, M. (2017) "Smartphones support smart labour", *Applied Economics Letters*, volume 24, issue 7, pages 467–471.

4 Microfinance

Microfinance tends to target rural regions that are not covered by city-based banks. Its original promise was to help transform rural livelihoods and to enable people – and often women in particular – to escape poverty. Often, the operational concept was that the route out of poverty would be through a series of increasingly substantial asset acquisitions. First a few ducks or chicken, then goats, then cows; or first a piece of land and then a larger piece of land; or first a sewing machine and then a workshop with multiple sewing machines. In almost all cases, microfinance products and services were and still are directly 'contributory', in the sense that using them comes at a price – be it through fees, interest payments or profit-sharing.

Microfinance has existed for a long time[1] but was relaunched in its modern form in 1976, in Bangladesh. From there, it spread across the wider Global South.[2] Modern microfinance started with micro*credit*: small loans to poor and predominantly rural people. It then expanded into saving schemes, which also came to include a modern and standardised version of group saving schemes (see Box 4.1). Then came leasing, pension and insurance products. The early providers had charitable aims. Commercial companies followed later. I do not know of research on the issue but have seen, in my own work, that the shift towards the latter was bolstered by a trend of donors supporting the development and roll-out of 'impact investments' that enabled fintech companies to enter the microfinance space.[3] Many of the products these companies provide are not yet commercially viable. For example, all micro-insurance initiatives I have seen or heard about still depend on external financial support.

Collectively, microfinance products and services are designed to move society in the direction of 'financial inclusion': the situation where accessible, useful, affordable and understandable financial products and services are widely available for whenever people and small companies

DOI: 10.4324/9781003451716-6

need them. Microfinance products offer people, prominently including rural poor people, the option of strengthening their livelihoods by borrowing money for a productive investment or by leasing an asset. Microfinance products also help smooth people's consumption (which may prevent hunger and hardship) by offering saving options to bridge income gaps and consumer loans for when times are tough. Pension products help avoid old age destitution; and crop and health insurance products avoid the need for distress sales. Provided that these products are designed to avoid debt traps, microfinance could also prevent poor people from falling into the hands of loan sharks (the evil 'other' in the minds of microfinance institutions and their donors) and, in the worst-case scenario, debt slavery.

Box 4.1 Village Savings and Loan Associations

Informal saving groups have existed for a long time: the oldest known system is the Korean 'Kye' system that may go back to the 9th century.[4] In 1991, an NGO called CARE developed a standardised version of such groups, piloted it in West Africa, and called its groups 'Village Savings and Loan Associations' (VSLAs). Like the informal saving groups, VSLAs help poor people build savings that they can use to invest in business or household assets, or spend at times of hardship or celebration. VSLAs are designed to ensure that they cause no friction, everything is clear to everybody, and no money disappears. Their results tend to be positive, and they don't suffer the indebtedness risks of microcredit.[5] Many other organisations adopted the same or similar models in their programmes – in fact I had come across VSLAs several times before I joined the CARE board of trustees, where I learned that they were all based on that 1991 CARE model.

Nowadays, many village-based saving and lending groups use mobile money or even online apps to facilitate their lending and saving, and for bookkeeping. This saves time and money, and is more secure than the traditional physical 'group saving box'. Mobile money has also come to represent *competition* to VSLAs: network operators increasingly offer saving facilities and people often store money in SIM cards as a no-interest-paying saving method (on which more in Chapter 5).

So the proposition was attractive and microfinance rapidly grew in turnover, geographical coverage and diversity of offer. It gained wide and ultimately global recognition.[6] Its meteoric rise culminated in a Nobel Peace Prize awarded in 2006 to Mohammad Yunus, the man behind that 1976 relaunch, and the Grameen Bank he had founded, "for their efforts to create economic and social development from below".[7] As the microfinance theme, and the focus on women borrowers therein, fit squarely in neoliberal Women's Economic Empowerment (WEE) and Women's Entrepreneurship Development (WED) approaches, micro-finance institutions tapped into substantial institutional donor funding.

However, by the time Mohammad Yunus and the Grameen Bank were awarded the Nobel Prize, an emerging body of research was already showing that microcredit was not actually living up to expectations. Three problems stood in the way: limited access, limited benefits, and the danger of causing harm. I cover them in turn, and in each case I explore how ICT may help address or otherwise affect the problem. Chapter 11 covers one more part of the story: the way ICT is *aggravating* the harm that microcredit inflicts on people by opening up rural regions for the global gambling industry.

Problem 1 of 3: Most poor rural groups do not have access to financial products. Notwithstanding decades of growth of global micro-finance, poor farmers still have less access to financial products than wealthier ones.[8] More generally, and at least until recently, there was "a huge mismatch between a limited microfinance menu and the demand for financial services by a highly heterogeneous population of more than two billion unbanked poor".[9] Much of the explanation is that poor people, and poor people who live far from the big cities in particular, are costly to reach, compared to the size of the loans, and that micro-financing them comes with high risks. I will first cover the risks and then the costs, and I will reflect on ICT-powered innovations that might increase the viability of microfinance operations that target poor and remote rural groups.

Risks. Conventional collateral does not work well for microfinance products, as poor people often do not have collateral that can easily be monetised (and poor women even less so, as land is often registered in the name of men). Microfinance institutions have developed alternatives. Some are based on cruelty. I once interviewed a woman who told me that "when I couldn't pay, they took my second sari to punish me, and now I only have what I'm wearing". Some lenders shame defaulters by yelling at them in public or forcing them to sit in front of their office. The more humane ones use the very widely replicated 'lending ladders' (which allow people to borrow ever-increasing sums of money, provided they have repaid all their previous loans) and group guarantees (where the group takes collective responsibility for its members' loans).[10] Some microfinance institutions use stored harvests as collateral.[11]

In three major ways, ICT reduces risks on the side of the lender.

1 *ICT facilitates individual profiling.* In countries where citizens have a digital ID (see below) and where this digital ID is used for 'know-your-customer' purposes (using systems that link personal data to people's digital ID), a digital credit history check has become an easy and low-cost way to reduce a lender's risks. Even more sophisticated are "novel methods of big data-driven credit scoring [which mean that] unbanked people are 'sorted' into behaviour-based risk categories as the basis for targeted marketing and risk pricing".[12] Such big data usage is reviled by ICT critics as abuse of data and corporate power.[13] However, it has an advantage as well because, in time, such practice may reduce the role of informal moneylenders, who managed to survive the advent of microfinance institutions and operate in parallel to them, even though the interest rates they charge tend to be far higher. So far, informal moneylenders have been able to maintain their position in part because their local presence gives them an informational advantage over microfinance institutions.[14] This advantage is smaller where personalised credit checks, via digital IDs or phone and internet usage data, are possible.

2 *ICT has created new forms of collateral and enables asset de-acti-vation in cases of default.* One example is Livestock Wealth, an app that enables investors to buy *physical* but not *fixed* assets, such as macadamia nut trees, oxen and pregnant cows. A farmer tends to these assets, and then buys them after the harvest, slaughter or birth of the calf. It is comparable to a loan, with the asset as the collateral, except that the ownership stays with the investor until the investment is generating money.[15] Another example is MKopa, "a connected asset financing platform that offers millions of under-banked customers access to life-enhancing products and services".[16] MKopa provides a pay-as-you go service for off-grid electricity, and is one of the companies that uses 'digital collateral': if people don't make their scheduled payments, the solar panels are simply switched off remotely.[17] For MKopa, this is cheap, as its 'lockout technology' does not require the physical removal of the asset, and it is easily reversible once payments have been made. For the user of a solar panel, a switch-off is painful without being crippling, which means that it provides a strong incentive not to default on

payments due, but it does so without the risk of a microfinance product leaving borrowers heavily indebted. Early evidence confirms that lockout technology works as intended: research on default rates for SHS, a Ugandan pay-as-you-go solar home system that has a similar switch-off system and uses its solar assets as digital collateral for school fee loans, found that, under otherwise identical circumstances, default rates for loans that use digital collateral are indeed significantly lower than for unsecured loans.[18] Livestock Wealth, MKopa and SHS are examples of a new group of ICT-driven products that keep risks manageable on the sides of both lenders and borrowers.

3   *ICT facilitates insurance that reduces repayment risks.* The scope of insurance products is widening every year, and they now cover health, life, assets and even communal property. One of the more recent market entrants is FundiFix, a maintenance service provider that was established in 2018. FundiFix insures against the breakdown of water pumps that are fitted with motion detectors – and the repair team is alerted to a potential breakdown if that detector signals that the pump has not been used for a while.[19]Agricultural risks tend to be particularly high. If the harvest fails or crop prices are low, borrowers may be unable to repay their loans because their harvests are of little value. Their group members cannot repay the loans on their behalf because *their* harvests do not have much value either, as crops typically fail regionally rather than just on an individual farm, and crop choices among peer farmers are likely to be similar.[20] This is where insurance has had its largest impact, because satellite imagery has made a combination of microcredit and index-based crop insurance an affordable and far less risky option for both lenders and borrowers. Some ICT critics do not agree that micro-insurance is taking off (as "demand for IBAI [index-based agricultural insurance] remains stubbornly low" because of the "unsuitability of the product"). They do not like the concept of IBAI either, as they see it as yet another way in which the agri-industry, with support from the donor community, is luring small-scale farmers into large agricultural value chains, and "can undermine the social relations essential to agricultural skilling".[21] These concerns were once valid but are now somewhat outdated. The early insurance products were crude but the more recent ones tend to be better suited to the localised needs of small-scale farmers; the demand for insurance products tends to increase over time, in regions where farmers have seen proof of the concept; and insurance products are not necessarily conditionalised on the

use of specific trading channels. The argument that micro-insurance may undermine social relations, is far-fetched. It simultaneously *under*estimates the rigour of social relations in many parts of the rural Global South, and *over*estimates the role of such relations in the case of harvest failure (which is likely to affect most people within a single localised social network). All in all: in case of well-designed insurance products, their advantages (less income variability, no need for distress sales) make them very worthwhile. I return to the issue of micro-insurance in more detail below, under "Problem 2 of 3".

Costs. Irrespective of loan size, a lending operation requires several steps. As steps take time and time costs money, each loan incurs a minimum cost. Bigger loans take more time, but not proportionally so, which means that small loans are relatively expensive to agree on and manage. Some of the costs are related to the time and effort it takes to get somewhere. Double the distance incurs much more than double the costs, because the general rule is that locations further from urban centres have worse and fewer roads, and more dispersed populations. The selection costs go up a little more if institutions restrict their products to certain groups, as microfinance programmes often do, such as people living with a disability or demobilised soldiers.

Traditional (i.e., non-digital) microfinance institutions limit their costs by operating in regions with high population densities. They also rely on local groups to do much of their group members' administration (and to collectively carry part of the lending risk). And they link up with NGOs that have better outreach capacity, are better able to find people who fit the lending criteria, and are willing to promote the uptake of microfinance products and take responsibility for some of the paperwork. Even so, microfinance organisations often either charge double-digit interest rates or require donor subsidies.

Today's digital options are reducing microfinance costs in ways that did not exist before. Digital "know-your-customer" options vastly reduce the complexities of due diligence procedures. As long as there is a 2G network to enable mobile money transfer, there is no longer a need to get actual money to remote locations. Similarly, there is no need for actual signatures, paper-based ladder tracking and face to face visits to groups in case of defaults (provided the group guarantee can be claimed digitally). Not all of this is common practice yet, but that is probably a matter of a mere few years and, like the risk-reducing effects of ICT innovations, this will help resolve the current microfinance access challenges.

Problem 2 of 3: Most people who use microfinance products do not end up with significantly stronger livelihoods (though micro-insurance might at least *protect* these livelihoods). Websites from microcredit institutions present unrepresentative success stories to let their potential sponsors know that "a loan from you can change someone's life" and that "people like José Livio Bravo are determined to work themselves out of poverty – and a loan of just £15 from you can help them get their business off the ground".[22] Evaluation reports will often present a somewhat more realistic picture, but only the strongest evaluators are able to withstand the positive bias pressure they are subject to. This is because organisations normally exert considerable effort to demonstrate success, whenever an evaluator is visiting, and it is difficult to avoid staging or to ensure a somewhat representative sampling method. In the case of microfinance programmes this leads to visits to very successful borrowers. Indeed, in the course of my evaluations I have had more encounters with people (women, mostly) who started with very little and managed to develop thriving rural shops, farms, and tailoring workshops, than encounters with people for whom microcredit had made no difference or worse. Similarly, I learned that, in focus group discussions, the more successful borrowers tend to sit in the front and talk more than the people who have benefited less or not at all, or who have faced problems because of the microloans they took.

The bias in more systematic microfinance research is less strong than the bias in evaluation reports, but it is not absent. Maren Duvendack and Philip Mader found, in a "systematic review of reviews" of micro-finance programmes, that "the more rigorous and lower-risk of bias studies are, the less likely they are to find effects" and that "this applies to both our reviewed meta-studies and to the underlying studies that constituted their evidence base".[23] In 2015, the *American Economic Journal* published a special issue that reported on carefully designed microfinance impact research that seemed to have avoided a positive bias to the fullest extent possible. The overriding finding from research conducted in Bosnia, Ethiopia, India, Mexico, Mongolia and Morocco (all except for India with a rural and entrepreneurial focus) was that microcredit is not "transformative in the sense of lifting people or communities out of poverty" but that it "does afford people more free-dom in their choices (e.g., of occupation) and the possibility of being more self-reliant".[24]

Sometimes, design tweaks within traditional microfinance programmes could improve their results. I have seen rural microcredit facilities that were ill-adapted to the agricultural cycle because their maximum grace periods ended before there was a sellable harvest, or that only offered credit suitable

The results of microfinance often disappoint

for single-year crops (which are not generally the higher-value crops). When interviewing borrowers, it sometimes struck me that more concerted efforts to improve financial literacy among borrowers could also help people choose, as people do not always fully understand what they have signed up to and may opt for inappropriate loans that leave them with unserviceable debts. But such design tweaks do not address the larger problems, which are unrelated to the microfinance products themselves. These problems are related to the limits of undiversified rural economies where markets are quickly saturated; the 'poverty trap' that poor people find themselves in, which means that a bit of extra money is unlikely to have lasting benefits; and the risks over which they have no control. A recent example I saw of just how powerless rural communities can be in the face of adversity was in the south of Ethiopia, in February 2020, where entire communities fruit-lessly tried to chase massive locust swarms out of their fields. Most ICT solutions amount to 'design tweaks' as well. As such, they may allow for results optimisation *within* microfinance's limited potential scope to improve livelihoods, by fine-tuning lending products and by offering better real-time coaching, for example, but they will not generally make microfinance transformational until these larger problems are resolved.

There is, however, one potential game changer, and this is the advent of radically new insurance products. These products potentially provide something akin to a social safety net[25] and, as such, may first and foremost help *protect* livelihoods – and indeed lives. The sense of security this creates may also improve people's well-being and encourage healthy entrepreneurial risk-taking, which may strengthen livelihoods. The impact of micro-insurance has not yet been adequately assessed but initial evidence suggests that its effects are positive. The two most common types of micro-insurance are health insurance and harvest insurance.

- *Health insurance.* ICT solutions have given health micro-insurance a boost. Such insurance may help to prevent distress sales when household members get sick, and may avoid the phenomenon of losing a relative after first losing all the household's assets to medical bills (a phenomenon that is sufficiently common for Arabic to have a saying that describes it: "موت وخراب ديار", "death and ruin"). It improves rural lives in regions with a functioning health care system.[26] In countries where health care systems require investments, PharmAccess has been trailblazing a set of interrelated products that address both the demand and the supply side of health care, by supporting phone-based micro-insurance schemes in Ghana, Kenya, Nigeria and Tanzania *and* providing digital loans to health clinics, with the aim of creating a positive spiral of ever-better health care facilities and ever-more health-care users. During a March 2021 meeting with the PharmAccess CEO, I was pleased to learn that the first direct competitor (i.e., a provider of a similar set of products) was about to enter the market. She presented this as bad news for PharmAccess, but it is good news for society, as it brings competition to the sector and confirms that her non-profit enterprise developed an idea with replicable potential.
- *Harvest insurance.* ICT – and specifically satellite imagery and localised meteorological data collection systems – opened the door to large-scale agricultural micro-insurance. This is an 'index insurance', which means it uses easily quantifiable and verifiable criteria that are specific to a *region*, rather than to a particular farm. This works better than insuring the specific harvests of individual farms, which would require costly inspections and would run the risk of fraud and of farmers not doing their utmost to achieve a good harvest. Index insurance pays out if the weather is too hot or cold, or there has been widespread flooding, or too little rain, or a cyclone or locust swarms, or if satellite imagery shows widespread crop failure in a farmer's wider region. Because an insurance pays out when (predefined) things go wrong, this may reduce people's harmful coping behaviours in case of a crop failure, such as distress sales and skipping meals, and risky behaviour like sex work, begging and illegal economic activities.

Another potential benefit of micro-insurance is that it might encourage more risk-taking entrepreneurial behaviour. Some researchers acknowledge these advantages but see them as short-term benefits that, in the long run, will be more than offset as agricultural micro-insurance facilities manage to lure small-scale farmers into modern forms of agriculture that will ultimately *add to* rather than *reduce* their exposure to climate risks.[27] That may be true (I do not know), but early research tentatively (because relatively little research has been done to date) confirms that, for now, the net effects of agricultural micro-insurance are positive:

> While the related empirical literature on the impacts of index insurance remains thin, it also underscores the promise that has motivated these initiatives. Households with index insurance reduce their dependence on detrimental coping strategies during severe shocks [...], increase investments in production, and, in some cases, make riskier production choices, which are all consistent with the mechanisms described by economic theory [...]. There is also evidence that insurance increases average total farm revenue (net of insurance premiums and indemnity payments) among Ghanaian farmers [...], and produces gains in both livestock productivity and child health for pastoralists in Kenya [...]. In addition, the peace of mind of having insurance coverage increases the subjective welfare of insured Ethiopian households, even in cases when there have been no indemnity payments.[28]

Notwithstanding their potential benefits, micro-insurance products sometimes face legal challenges. In the sample of one of my more recent donor evaluations, for example, there was a rural micro-insurance company in Cambodia that had to halt operations because of an objection from the country's banking regulator. A common problem is that index insurance does not fit legal definitions of 'insurance', because payments are linked to *assumed* losses rather than *actual* losses, and as such one could argue that "indexed products are, in fact, weather derivatives for farmers, not insurance that guarantees recompense for losses. They presume that astute farmers can and should manage environmental risks through offsetting bets on financial products".[29] In addition, and notwithstanding massive investments to market index insurance among small-scale farmers,[30] micro-insurance products often face challenges with their take-up.[31] Some fintech companies, including the otherwise mighty M-Pesa, launched micro-insurance products but discontinued them as they failed to gain traction.[32] There are many other challenges still to overcome.[33]

In spite of these challenges and concerns, the sector is expanding rapidly. Many governments and ODA-financed stakeholders support this expansion process. They give insurance providers grants, provide subsidised venture capital for rural product development and roll-out, subsidise micro-insurance premiums to encourage subsequent uptake of these products, and re-insure insurers against large-scale calamities.[34] The result is that the design of micro-insurance continues to evolve[35] while, in parallel, there are already schemes with hundreds of thousands of subscribers – national index insurance programmes in India are collectively reaching over 30 million people.[36]

While existing micro-insurance schemes have been fine-tuning their designs and developing better links to a range of microfinance products, other companies are developing competing products. There are promising experiments with guaranteed credit lines that become available once adversity kicks in (so that people know they are going to be able to get credit if they encounter problems), for example.[37] There is also the beginning of peer-to-peer 'weather immunity tokens', where investors and farmers both contribute to a fund that pays out to either one of them after the harvest season, depending on what the weather has been like.[38]

Despite this ongoing growth in both the volume and the diversity of microfinance products, the overall results of microfinance operations to date have been disappointing, and sometimes microfinance (and micro-*credit* in particular) causes people harm. This is the third and last of the problems that is affected by ICT.

Problem 3 of 3: Microcredit causes people harm. In microcredit programmes in Ethiopia, Bangladesh, Mexico and Morocco, only 31%, 20%, 19% and 17% respectively of the people who *could* take a loan actually *did* take one.[39] Part of the explanation is that many people find loans too risky and therefore unattractive. Their reluctance to participate is called 'risk rationing',[40] and this is a sensible attitude for consumer and investment loans alike, as both types can cause serious problems.

- Consumer loans are meant to smooth consumption by providing money in times of hardship, to be repaid in times of relative plenty. In reality, such loans are often used to reduce hardship today – but without reasonable expectations of better times tomorrow. In such cases, and over a period of time, the net result is that a household's total consumption is lower than it would be without microcredit, because of the fees and interest payments. People end up "eating leaves and salt"[41] in order to repay their loans.

- Investment loans are meant to increase income, but the obvious risk is that the results of investments fall short of expectations. If this happens, gains may be insufficient to cover the interest payments, and repayments may require selling more assets than were bought with the loan. Borrowing costs may further increase if households or businesses then incur new debts in order to repay old ones, or if their repayments are late and fines and interest-on-interest are added to the amount originally owed (a problem that does not exist in Islamic microfinance). Or households or companies may end up defaulting on their loans altogether, and lose their collateral or standing in the community. In the worst-case scenario, this leads to death. In India and Bangladesh in particular, there are many documented cases of people's inability to repay their loans causing such despair that they chose to end their lives.[42] A string of such microfinance tragedies led Jairam Ramesh, when he was the Rural Development Minister of India in 2012, to say that:

> nothing can stop an idea whose time has gone. And microfinance [he means microcredit in particular] is a discredited model. It has raised more questions that it has answered. To think that we are going to alleviate poverty is a tall claim. Microfinance has promised more than it has actually delivered, created more problems than it actually solved and continues to promise much more than what it actually puts on the ground.[43]

ICT can worsen the problem. On the one hand, the digital know-your-customer checks (or, in countries where these do not yet exist, the less thorough method of microfinance institutions linking their databases) can help, as such checks not only reduce the lender's risk but also protect borrowers against borrowing beyond their repayment capacity. On the other hand, mobile credit services make it easier for people to get heavily indebted. These services are often instant, and they are therefore prone to borrowing on the basis of impulse rather than carefully made choices. In addition, annualised interest rates of mobile credit sometimes go into several hundred percent.[44]

—

All in all, ICT applications are working towards optimising microfinance products and services. These applications are often developed by companies that do not see themselves as microfinance institutions, such as mobile app developers and companies that transfer migrants' remittances (see the next chapter). They bring fresh thinking to the microfinance sector and the result is that, every year, microfinance providers penetrate more remote and poorer rural areas, and offer more appropriate and appealing products. On the lenders' side, they are unlocking new sources of financing such as peer-to-peer lending (see Box 4.2), and they are reducing risks which, in time, is likely to reduce both the costs of microfinance and the harm it sometimes causes. The pace of change will accelerate because ICT makes every next step a little easier, and only a few years from now a wide range of products and services will be available wherever there is a minimum degree of stability.

Box 4.2 Online peer-to-peer lending

The first online peer-to-peer lending platforms started in 2005, in Canada and the UK. These platforms match individuals who have money to invest with borrowers who struggle to get loans through conventional banking channels. The idea caught on and the sector now issues billions of dollars in loans every month. It is a volatile sector. New models are being tried all the time and, in China in particular, many platforms suspend operations soon after they have started.

Most platforms are commercial and lend only within a single country, but there are also some non-profit crowdfunding platforms that link lenders from anywhere to poor and often rural borrowers in the Global South. These charitable platforms cover their costs through donations and government grants, and tend to charge little or no interest or fees (though their intermediaries generally do, and not necessarily low ones).[45] Some of these non-profit platforms, such as kiva.org and lendwithcare.org, have become large already, continue to grow more, and seem robust.

Such charitable peer-to-peer lending platforms face many of the same challenges as other forms of microcredit, but they have managed to generate a new flow of money. Moreover, the people behind them often do not originate from the microfinance sector and may well bring innovations that nobody has thought of yet.

But however wide its reach and however diverse its products and services, the microfinance sector is unlikely to be truly transformational until illiteracy, malnutrition, discrimination, bureaucracy, corruption and other challenges that keep people poor are resolved. Where this happens, the next generation of financial products has the potential to reduce short-term hardship and improve long-term prospects. Even then, microfinance will not help the very poorest people, as they need grants rather than loans, as well as other forms of support (see Box 5.2, in the next chapter). It may also not benefit everyone equally and may reinforce existing forms of discrimination as "algorithms sifting through data about mobile phone or social media users may [...] determine that a person's ethnic or religious identity, geographical location or social relations provide proxies for credit worthiness and may facilitate finance accordingly".[46] Lastly, microfinance will pose a danger to rural people who gain access, through their mobile phones and often for the first time, to gambling – which is an issue I will return to in Chapter 11.

Notes

1 According to James Trager, the first documented agricultural leasing products date back 2,800 years. Trager, J. (1995) *The food chronology*, Henry Holt, section on 800 BC. For case studies from 18th century Ireland and Germany, and a passing reference to 15th century Nigeria, see Seibel, H.D. (2003) "History matters in microfinance", *Working Paper Number 2003/5*, Universität zu Köln, Arbeitsstelle für Entwicklungsländerforschung.

2 For a brief outline of the model's Bangladeshi origins and its subsequent spread, see Zainuddin, M. and Yasin, I.M. (2020) "Resurgence of an ancient idea? A study on the history of microfinance", *FIIB Business Review*, volume 9, issue 2, pages 78–84.

3 Impact investments are investments that may be made by commercial companies for commercial reasons, but that have a development angle to them. Fintech companies are companies that produce ICT-enabled financial products and services.

4 Frits Bouman describes the history of saving groups throughout the world. See Bouman, F.J.A. (1995) "ROSCA: on the origin of the species", *Savings and Development*, volume 19, number 2, pages 117–148. (ROSCA stands for "rotating savings and credit association".)

5 There are many case study publications on the issue, focusing mostly on African countries, and the findings are nearly always broadly positive. E.g., Ksoll, C. *et al* (2016) "Impact of Village Savings and Loan Associations: evidence from a cluster randomized trial", *Journal of Development Economics*, volume 120, pages 70–85.

6 This, for example, is a 1997 statement from the Prime Minister of Bangladesh: "In our careful assessment, meeting the credit needs of the poor is one of the most effective ways to fight exploitation and poverty. I believe that this campaign will become one of the great humanitarian movements of history. This campaign will allow the world's poorest people to free themselves from the bondage of poverty and deprivation to bloom to their fullest potentials to the benefit of all – rich and poor". Quoted in Banerjee, A., Karlan, D. and Zinman, J. (2015) "Six randomized evaluations of microcredit: introduction and further steps", *American Economic Journal: Applied Economics*, volume 7, number 1, 21 pages, with the quotation in footnote 1.

7 Announcement of the 2006 Nobel Peace Prize to Muhammad Yunus and Grameen Bank, presented by Ole Danbolt Mjøs, Chairman of the Norwegian Nobel Committee, on 13 October 2006; see nobelprize.org/prizes/peace/2006/prize-announcement.

8 See, e.g., Ibrahim, S.S. and Aliero, H.M. (2012) "An analysis of farmers' access to formal credit in the rural areas of Nigeria", *African Journal of Agricultural Research*, volume 7, issue 47, pages 6249–6253; Akudugu, M. A., Egyir, I.S. and Mensah-Bonsu, A. (2009) "Women farmers' access to credit from rural banks in Ghana", *Agricultural Finance Review*, volume 69, issue 3, pages 284–299; Chisasa, J. and Makina, D. (July 2012) "Trends in credit to smallholder farmers in South Africa", *International Business & Economics Research Journal*, volume 11, number 7, pages 771–783; and Saqib, S.E., Ahmad, M.M. and Panezai, S. (2016) "Landholding size and farmers' access to credit and its utilisation in Pakistan", *Development in Practice*, volume 26, issue 8, pages 1060–1071.

9 Armendáriz, B. and Labie, M. editors (2011) *The handbook of microfinance*, World Scientific, with the quotation from page 5.

10 This arrangement is sometimes criticised because it transfers the default risk from the microcredit institution to groups of poor people (and generally poor women in particular), but this is unfair criticism because the risk is much *reduced* through this transfer, as the group has a large informational advantage over the microcredit institution, meaning that it is far better able to judge the members' ability to repay their loans. See Stiglitz, J.E. (September 1990) "Peer monitoring and credit markets", *The World Bank Economic Review*, volume 4, number 3 (a special issue on the issue of "imperfect information and rural credit markets"), pages 351–366.

11 These loans are typically meant to enable farmers to store their harvest until the lean season, when crop prices are higher. See, e.g., Channa, H. *et al* (November 2018) *Helping smallholder farmers make the most of maize through harvest loans and storage technology: insights from a randomized control trial in Tanzania*, Purdue University; and Sagbo, N.S.M. and Kusunose, Y. (July 2021) "Does experience with agricultural loans improve farmers' well-being? Evidence from Benin", *Agricultural Finance Review*, volume 81, number 4, pages 503–519 (look for the word 'warrantage').

12 Brooks, S. (2021) "Configuring the digital farmer: a nudge world in the making?", *Economy and Society*, volume 50, issue 3, pages 374–396, with the quotation on page 385.

13 *Ibid*.

14 This informational advantage enables them to set interest rates that reflect default risk levels, which helps explain why informal moneylenders sometimes thrive while the formal institutions struggle with losses. Another advantage for informal moneylenders is their ability to ensure that borrowers use the loans productively. See the introductory pages of Stiglitz, J.E. (September 1990) "Peer monitoring and credit markets", *The World Bank Economic Review*, volume 4, number 3, pages 351–366.

15 See livestockwealth.com or, for a brief description of the way physical assets serve as collateral in a Livestock Wealth contract, see page 23 of Aguera, P. *et al* (June 2020) *Paving the way towards digitalising agriculture in South Africa*, Research ICT Africa.

16 M-kopa.com, with the quotation from the 'about' section.

17 Mutongwa, S.M. and Abeka, S. (2020) "Implementation of MKopa solar services for poverty eradication", *Journal of Scientific and Engineering Research*, volume 7, number 1, pages 76–87.

18 By 19%, according to Gertler, P., Green, B. and Wolfram, C. (April 2021, revised May 2021) "Digital collateral", *NBER Working Paper 28724*, National Bureau of Economic Research.

19 Dahmm, H. (undated but 2018) "Handpump data improves water access", *Case study by SDSN TReNDS*, Global Partnership for Sustainable Development Data. Like most micro-insurance initiatives, FundiFix is only very partially covered by community contributions.

20 I have seen passing references to harvest *prospects* being used as collateral. Both informal lenders and formal microfinance institutions sometimes do this – see, respectively, Zahana (undated) *Microcredit or community lending systems for rural Madagascar*, Association Zahana; and Konare, K. (2001) *Challenges to agricultural finances in Mali*, Michigan State University. However, I have not come across this myself and could not find research on it. The rarity of this type of collateral is probably the consequence of a paradox: if a lender needs to claim collateral because of a default on an agricultural loan, it is probably because of harvest failure, in which case the particular collateral of a prospected harvest cannot be claimed.

21 Brooks, S. (2021) "Configuring the digital farmer: a nudge world in the making?", *Economy and Society*, volume 50, issue 3, pages 374–396, with the quotations on page 383 and 388.

22 This is the opening text of lendwithcare.org, visited on 29 June 2021, with a picture of a man tending to his trees and a button with the invitation to "learn more".

23 Duvendack, M. and Mader, P. (2019) "Impact of financial inclusion in low- and middle-income countries: a systematic review of reviews", *3ie Systematic Review*, number 42, 3ie, with the quotations from page 41.

24 Banerjee, A., Karlan, D. and Zinman, J. (2015) "Six randomized evaluations of microcredit: introduction and further steps", *American Economic Journal: Applied Economics*, volume 7, number 1, 21 pages, with the quotation from page 12.

25 Except that people typically have to pay for insurance while social assistance programmes are non-contributory.

26 For example, Zhonkun Zhu and his colleagues concluded that "purchasing medical insurance helps improve farmers' psychological health because insurance relaxes potential anxiety over the risk of illness and ensure life security [*sic*]." Zhu, Z., Ma, W. and Leng, C. (2020) "ICT adoption, individual income and psychological health of rural farmers in China", *Applied Research on Quality of Life*, 21 pages, with the quotation from page 15.

27 Isakson, S.R. (October 2015) "Derivatives for development? Small-farmer vulnerability and the financialization of climate risk management", *Journal of Agrarian Change*, volume 15, number 4, pages 569–580. According to Ryan Isakson, the very *purpose* of micro-insurance is less to protect farmers than it is "to integrate smallholder farmers into agri-food value chains where debt and dependence upon commercial inputs reorient agricultural production and the associated distribution of value [and that] the construction of an accommodating insurance culture is key to this end" (the quotation is from page 574).

28 Jensen, N. and Barrett, C. (2017) "Agricultural index insurance for development", *Applied Economic Perspectives and Policy*, volume 39, number 2, pages 199–219, with the quotation on pages 200–201. The ellipses replace the references that substantiate the claims. For a study about coping behaviour in tough times, see Janzen, S.A. and Carter, M.R. (2019) "After the drought: the impact of micro-insurance on consumption smoothing and asset protection", *American Journal of Agricultural Economics*, volume 101, issue 3, pages 651–671.

29 Isakson, S.R. (October 2015) "Derivatives for development? Small-farmer vulnerability and the financialization of climate risk management", *Journal of Agrarian Change*, volume 15, number 4, pages 569–580, with the quotation from page 570. This is an exaggerated statement because the idea is that the pay-out criteria and harvest failure are closely related, but it is obviously true that "the effectiveness of index insurance as a risk mitigation tool depends on how positively correlated farm-yield losses are with the weather index [and that] unless the index is based on a weather variable that is the dominant cause of loss in the region, [the] basis risk [i.e., the risk of a disconnect between an index estimate and a farmer's actual experience] will be unacceptably high". Nair, R. (February 2010) "Crop insurance in India: changes and challenges", *Economic & Political Weekly*, volume xlv, number 6, pages 19–22.

30 For an account of efforts in India, where index insurance has grown most rapidly, see Da Costa, D. (2013) "The 'rule of experts' in making a dynamic micro-insurance industry in India", *The Journal of Peasant Studies*, volume 40, issue 5, pages 845–865.

31 See, e.g., Budhathoki, N.K. *et al* (2019) "Farmers' interest and willingness-to-pay for index-based crop insurance in the lowlands of Nepal", *Land Use Policy*, volume 85, pages 1–10.

32 M-Pesa's short-lived health micro-insurance is mentioned on page 143 of Aron, J. (August 2018) "Mobile money and the economy: a review of the evidence", *The World Bank Research Observer*, volume 33, number 2, pages 135–188. (M-Pesa also serves as the platform for a more successful micro-insurance facility called Kilimo-Salama, which provides agricultural insurance products.)

33 For a review of current challenges and likely future developments in fields such as quality standards and impact-based targeting, see Jensen, N. and Barrett, C. (2017) "Agricultural index insurance for development", *Applied Economic Perspectives and Policy*, volume 39, number 2, pages 199–219.

34 The World Bank's Global Index Insurance Facility, or GIIF, tops the list of ODA-financers, and the Government of India is probably the world's largest subsidiser among national governments. For a discussion of the politics of subsidies (mostly in the case of India), see Da Costa, D. (2013) "The 'rule of experts' in making a dynamic micro-insurance industry in India", *The Journal of Peasant Studies*, volume 40, issue 5, pages 845–865, with the relevant section on pages 852–855.

35 For an example of sophisticated design deliberations, see Shirsath, P. *et al* (2019) "Designing weather index insurance of crops for the increased satisfaction of farmers, industry and the government", *Climate Risk Management*, volume 25, 12 pages.

36 For more on micro-insurance programmes in India and other countries, see Greatrex, H. *et al* (2015) "Scaling up index insurance for smallholder farmers: recent evidence and insights", *CCAFS Report*, number 14.

37 Lane, G. (November 2020) *Credit lines as insurance: evidence from Bangladesh*, American University.

38 Jha, S., Andre, B. and Jha, O. (2018) *ARBOL: smart contract weather risk protection for agriculture*. Note that this is a promotional paper. To my knowledge, no academic research has been conducted yet on the issue of peer-to-peer micro-insurance, and I do not know if these tokens got off the ground.

39 The percentages for Ethiopia, Mexico, Morocco are discussed in the corresponding country studies covered in a 2015 special microfinance issue of the *American Economic Journal: Applied Economics*, volume 7, number 1, and summarised on page 10 of the introductory paper of that special issue. The figure for Bangladesh is from Hossain, M. *et al* (2019) "Agricultural microcredit for tenant farmers: evidence from a field experiment in Bangladesh", *American Journal of Agricultural Economics*, volume 101, issue 3, pages 692–709.

40 Widespread risk rationing also means that efforts to formalise land rights and security of tenure may not increase microcredit uptake, even though this is often stated to be one of the objectives of such efforts, as formalised land ownership open possibilities for collateralised lending.

41 This quotation is part of a longer text from Banerjee, S.B. and Jackson, L. (2017) "Microfinance and the business of poverty reduction: critical perspectives from rural Bangladesh", *Human Relations*, volume 70, issue 1, pages 63–91, with the quotation from page 75.

42 An experiment with micro-borrowers in rural India showed that group-based microfinance sometimes comes with excessive peer pressure and peer punishment. See Czura, K. (2015) "Pay, peek, punish? Repayment, information acquisition and punishment in a microcredit lab-in-the-field experiment", *Journal of Development Economics*, volume 117, pages 119–133. This paper hyperlinks to a few newspaper articles about the problem of microcredit-related suicides.

43 Quoted in several papers; e.g., Banerjee, S.B. and Jackson, L. (2017) "Microfinance and the business of poverty reduction: critical perspectives from rural Bangladesh", *Human Relations*, volume 70, issue 1, pages 63–91, with the quotation from page 64.

44 For a concise overview of the problems and possible ways forward, see The Economist (17 November 2018) *Borrowing by mobile phone gets some poor people into trouble*, The Economist.

45 The only figure I have seen is from Kiva, from 2014: "Kiva's Field Partners […] usually charge interest […] and the average portfolio yield of our microfinance partners is ~35 percent" (nextbillion.net/kiva-responds, last seen on 4 December 2022).

46 Mann, L. (2017) "Left to other peoples' devices? A political economy perspective on the big data revolution in development", *Development and Change*, volume 49, issue 1, pages 3–36, with the quotation from page 17.

5 Remittances and social assistance

As a no- or low-cost, quick and secure payment method that works across large distances, mobile money rapidly gained a foothold around the world. Over the past decade, the number of registered accounts, the number of active accounts, and the number and value of transactions have all rapidly increased in all regions of the world.[1] Even the 2020 pandemic has not halted its growth.[2] Some research concludes that mobile money contributes to poverty reduction,[3] but this contribution is not easy to isolate. A meta-review of studies that aimed to do so found that most research conclusions on the mobile money effects on poverty, inequality and other development dimensions do not hold up to scrutiny[4] – notwithstanding mobile money's obvious facilitating role in financing and transactions, and in the creation of opportunities through e-commerce.

Two key areas in which the importance of virtual money for many millions of people is largely undisputed are remittances and social assistance. Both streams are sizeable and flow largely from urban to rural regions. Together they often provide more than half a rural household's income. Both streams are largely 'non-contributory': they are transfers that are not the result of people *buying* an entitlement. This sets them apart from microfinance products and services. In both cases, mobile money has extended the reach, simplified the transfer process and reduced the costs of transfers. This has changed the behaviour of senders and their rural recipients. This chapter covers these two streams of income in turn.

Remittances

Virtual money providers initially targeted migrants – and those migrants provided a massive client group. Some 3.5% of the world's population are international migrants – an estimated 272 million people

DOI: 10.4324/9781003451716-7

in total.[5] The World Bank estimates that international remittances sent to low- and middle-income countries of origin amounted to some US $530 billion in 2018.[6] This estimate does not include cash informally crossing borders. It also does not include remittances that internal migrants send from the city to the village – which is where mobile money first gained traction. On average and per person, internal migrants do not transfer as much as international migrants do. However, the number of internal migrants far exceeds the number of international ones: there are roughly as many internal migrants in China (an estimated 274 million)[7] as there are international migrants worldwide, and there are an estimated 139 million internal migrants in India.[8]

Economic migrants are often rural, single, young, able-bodied men and (a little less commonly) women. They have predominantly rural backgrounds, and often migrate to cities, within their own country or abroad. In parallel, there is intra-rural seasonal migration and cross-border rural-to-rural migration (Egyptians in Jordan often work on farms, for example). Whatever their destination, these migrants' departure from the region of origin has a major impact on rural lives and livelihoods there. It affects the local labour market – either because migrants leave labour and skill gaps (possibly causing a move towards less labour-intensive crops, for example) or because their departure increases the work opportunities for those who stay behind, as these opportunities are shared among fewer people. It affects gender patterns because, when either men or women are absent,[9] people from the other sex have to perform duties that were not traditionally theirs. But it is *remittances* that change rural lives and livelihoods most profoundly. These flows are not directly contributory: often households and wider social groups invested in a person's migration, and expect to be rewarded, but the link is not normally direct and the amount expected not normally pre-set. Instead, remittance flows tend to continue for a long time, are largely resistant to recessions[10] (or at least this was the case until the 2020 pandemic[11]) and are *so* important that the UN General Assembly recently made every 16th June the *International Day of Family Remittances* because "in many developing countries international remittances constitute an important source of income for poor families [and] half will reach rural areas".[12]

Research across the three regions of the Global South found that the overall economic effects of remittances are positive, and more so where local markets function well (because of the multiplier effect) and where the business environment is sufficiently attractive for recipients to *invest* rather than just *spend* their savings.

The multiplier effect of remittances

- In Latin America and the Caribbean, research that used a 10-country panel data set concluded that "remittances [...] have increased growth and reduced inequality and poverty".[13]
- In Asia and the Pacific, research using panel data for 24 countries concluded that remittances contributed to poverty reduction – especially through their direct effects – *and* benefited overall economic growth.[14]
- Research using data from 34 countries in Africa concluded that remittances significantly reduced poverty.[15]

Because these remittances are of such importance to poverty alleviation, one of the Sustainable Development Goal targets for 2030 is to "reduce to less than 3 per cent the transaction costs of migrant remittances and eliminate remittance corridors with costs higher than 5 per cent".[16] This is not the first target related to the cost of international remittances (in 2009, the G8 agreed, unsuccessfully, on the "5X5 objective" of reducing international remittance prices from 10% to 5% between 2009 and 2014), but rapidly evolving ICT applications mean that this time these targets seem within reach.

For in-country transfers, transaction costs are generally already lower than the current 3% target, because of mobile money. Many mobile money options were designed primarily for the purpose of facilitating internal remittances, and were marketed as such. Remittances are particularly important in the rural economy of the Philippines, and the Philippines were one of the very first countries in which mobile money products took off (soon after a product named Smart Money was launched in 2001). Although the range of uses for mobile money has widened since those early days, research conducted eight years later – in 2009 – still found that "low-income Filipinos are *primarily* using mobile money to send and receive domestic remittances".[17]

Nearly all regions in the world now have at least adequate 2G coverage, and where this is the case mobile phones are often the most important transfer channel for in-country remittances, including in Africa,[18] where Kenya's M-Pesa led the way. After the failure of its initial idea of facilitating microfinance, M-Pesa's "key proposition" at the start of its launch year (2007) was to "Send Money Home".[19] This was a key service for migrants who faced difficulties sending their remittances back to the village, and the uptake was instant. Other companies followed, often using a similar text-based modality that makes easy electronic transfers through 2G networks. People use M-Pesa and equivalent 2G⁺ providers *a lot* – and although I have not seen post-pandemic figures, I assume that the proportion of remittances sent in the form of mobile money will have further increased when the pandemic restricted travel, as this complicated the physical carrying of money from the city to the village.[20] The option of mobile money transfers might have encouraged out-migration from villages[21] and, more unambiguously, it changed the effects of remittances in two distinct but related ways.

The first effect of sending remittances as mobile money is that it smooths consumption. This happens in three ways:

- The lower cost-per-transaction leads to migrants transferring frequent small amounts instead of less frequent but larger amounts.[22] On the side of the receiver, this reduces the perceived risk of being without an income, which in turn helps stabilise expenditures. I have not seen research coming to this conclusion in the context of mobile payments, but there is evidence within the social assistance literature that regularity and predictability contribute to consumption smoothing.[23]

- ICT reinforces the phenomenon that "remittances received by smallholder farm households are often found to be higher in years with below average agricultural incomes".[24] This phenomenon itself is not new. Remittances always partly depend on needs, and in fact sometimes the flow reverses and rural household members make transfers to their urban members. This happened in 2008, for example, during Kenya's post-election crisis, when urban migrants needed to escape the threat of violence.[25] However, mobile communication and the ability to transfer small amounts at high frequency without losing a large proportion to transaction fees have made the link much more direct and instant. The difference is significant: in India, a big-data piece of research that compared remittance channels within and across districts found that:

> mobile money can provide an insurance against shocks [and that] while rainfall shocks have a significant negative impact on economic activity, proxied by night-time lights [...] this effect is partially mitigated in districts that use mobile money. Specifically a 10 percent increase in mobile money use in districts hit by a rainfall shock reduces the negative effect of the shock by 3 percent.[26]

- Mobile network operators increasingly offer remittances-linked financial products (mostly saving facilities, insurance products and loans) and, even without any profit-driven encouragement, people often store money in SIM cards as a no-interest-paying saving method.[27] Although there is resentment and resistance to this 'financialisation' of remittances, in the literature and in practice,[28] it does widen the variety of saving mechanisms and this has a risk-reducing effect and allows for a higher level of consumption when incomes are temporarily low.

The second effect of sending remittances as mobile money is that it increases the sum total of transfers – at least in some African countries. Research in Kenya,[29] Ghana[30] and Uganda[31] suggests that these cheap mobile money transfers led to larger overall sums to be transferred over time. (I have not seen Asia-focused research on the topic.)

Although the advantages are substantial, mobile phone transfers are currently still mostly a vehicle for *in-country* transfers of remittances. International mobile money transfers are significantly cheaper than other forms of international transfers,[32] but the proportion of remittances transferred through mobile money transfers is still small (though there are exceptions).[33] A World Bank blog that sought to explain why the SDG target on the costs of international remittances was not close to being achieved merely mentioned mobile phone transfers in passing:

> There has been a rapid increase in mobile-phone and internet-based technology solutions for providing remittance services, and more recently, in block-chain-based applications. However, such fintech solutions almost exclusively rely on banks to provide the know-your-customer compliance requirements, thus limiting themselves to banked customers and leaving out a vast majority of unbanked customers who may be encouraged to use informal remittance channels.[34]

For progress to accelerate in the next few years, cross-border mobile money transfers need to become as common as in-country mobile money transfers are today.[35] That this is not currently the case is mainly because the rules and regulations around anti-money laundering and countering the financing of terrorism are creating obstacles. These rules and regulations are based on guidance provided by the influential FATF – the Financial Action Task Force.[36] (For reasons that I do not understand, this is never mentioned in publications that seek to explain the discrepancy between the growth patterns of in-country and cross-border mobile money transfers of remittances.) It is hard to predict how this will evolve in the next few years.

On the one hand, there is some largely unsubstantiated fearmongering, such as in a recent chapter on the "Impact of mobile money on financial crime, money laundering, and terrorism financing", which says that:

> With respect to the financing of the Somali-based terrorist group, Al-Shabaab, a report by the US Bureau for International Narcotics and Law Enforcement Affairs published in 2015 established that the main financing for terror activities is by mobile money and Somali's hawalas.[37]

As far as I am aware, no published evidence of this or any other link between mobile money and terrorism exists, and the report from the US Bureau for International Narcotics and Law Enforcement Affairs referenced in this quotation does not say anything of the sort. This 2015 report lists a number of other sources and channels of Al-Shabaab financing, but not mobile money,[38] and its section on India argues that India

> should consider further facilitating [...] mobile banking [because an] increase in lawful, accessible services would allow broader financial inclusion of legitimate individuals and entities and reduce overall AML/CFT [anti-money laundering/combating the financing of terrorism] vulnerabilities by shrinking the informal network, particularly in the rural sector.[39]

However, even if the evidence base of such claims is thin or absent, every new claim of such abuse provides a strong incentive, to FATF and national governments, to restrict international mobile money transfers.[40]

On the other hand, the advantages of low-cost mobile phone transfers are considerable, in terms of both costs and convenience, and blockchain technology may further reduce these costs.[41] In addition, mobile phone companies are actively marketing their individual products,[42] and collectively they are promoting the mobile money channel.[43] A few donors are supporting these advocacy efforts.[44] The various lobbying and advocacy efforts point out that mobile money options drive down the costs of other forms of international remittance transfers,[45] and they often refer to SDG target 10.c to bolster their pro-mobile money messages.[46] Their efforts, perhaps supported by lobbying on the part of diaspora organisations and their supporters, may well achieve something akin to the advocacy victory against FATF's 2001 "Recommendation 8", which used to hinder cross-border operations of civil society organisations (CSOs, see Box 5.1).

Box 5.1 An advocacy victory in relation to international transfers

During the 2018–2019 Independent Commission for Aid Impact (ICAI) review on DFID's partnerships with civil society organisations (CSOs)[47] we found that, for years, governments had used the 2001 "Recommendations on Terrorist Financing" of the Financial Action Task Force to clamp down on civil society organisations.[48] The key problem was the original 2001 formulation of FATF's "Recommendation 8", which asked countries to "review the adequacy of laws and regulations that relate to entities that can be abused for the financing of terrorism" and which noted that "non-profit organisations are particularly vulnerable, and countries should ensure that they cannot be misused".[49] As with the restrictions on international mobile money transfers, the drawbacks of this expectation of guilt far outweighed the advantages.

After a serious lobbying effort on the side of CSOs (after a while supported by DFID and a few other donors), the recommendation changed in June 2016. The new recommendation (updated a few times since, most recently in October 2021) does not consider all non-profit organisations to be "particularly vulnerable" and instead refers to "non-profit organisations which the country has identified as being vulnerable to terrorist financing abuse" and asks for "focused and proportionate measures".[50]

Social assistance

Social assistance is the provision of non-contributory transfers. These are provided in cash or in kind, sometimes by CSOs but more generally by governments, and they are targeted at poor and vulnerable people. School meals, for example, are part of social assistance, but general fuel or food subsidies are not because they lack specific targeting.

Developing countries invest heavily in social assistance – an average of 1.5% of their GDP.[51] This spending takes different forms. Public works programmes, fee waivers, food support and cash transfers all potentially qualify as social assistance. However, they are not all equal. Most research concludes that, where functioning markets exist, cash transfers are superior to other types of social assistance. In the ICAI review on "The effects of DFID's cash transfer programmes on poverty and vulnerability", my findings in relation to DFID's portfolio broadly aligned with other research and concluded that:

- Cash transfers did not live up to *all* of DFID's stated ambitions. They did not commonly and by themselves strengthen health, nutrition, resilience and children's learning, even though programme documentation regularly made such claims. However:
- The programmes in DFID's cash transfer portfolio *did* increase consumption, and they empowered recipients as they supported people without making assumptions about their needs. In some cases, cash transfers served as a useful part of multifaceted poverty reduction approaches (see Box 5.2). Cash transfers often reduced negative coping behaviours and intra-household inequalities. Overhead costs of cash transfer programmes were relatively low, and the transfers were easy to monitor and therefore difficult to corrupt, provided that the initial targeting was sound and based on needs and vulnerabilities rather than loyalties. Better targeting and timeliness of disbursements, and more appropriate – generally larger – transfer sizes, would further strengthen these effects. Even without these improvements, the effects of DFID's investments were such that this ICAI review recommended that "DFID should consider options for scaling up contributions to cash transfer programmes where there is evidence of national government commitment to improving value for money, expanding coverage and ensuring future financial sustainability".[52]

Box 5.2 BRAC's poverty graduation programme

A rural development organisation named BRAC started its 'poverty graduation programme' in Bangladesh in 2002, inspired by its observation that BRAC's work did not manage to be truly useful to the poorest rural people.

This programme begins with a very careful participatory process that identifies a community's poorest and most vulnerable members. Once BRAC knows who to target, it combines relatively small cash transfers with support such as asset grants (often some sort of livestock), animal vaccinations, food supplements, entrepreneurial and other training, and hygiene awareness raising. The results are good, and in one of the ICAI reviews I concluded that – in part because the weekly cash transfers – it prevented distress asset sales when shocks occurred:

> There is strong evidence that [BRAC's poverty graduation pro-
> gramme] has enabled a large majority of beneficiaries to achieve
> substantial improvements in their socioeconomic status. The
> chances of these mixed interventions achieving results that
> continue after households have exited the programme are much
> stronger than for pure cash transfers. Furthermore, the results
> have shown an ability to survive climate shocks.[53]
>
> The model may not be replicable everywhere, and may perform less
> well in times of economic downturn. It is also costlier than standard
> cash transfers, as it requires close monitoring and action on multiple
> fronts. But it is the only model I know of that does genuinely and
> sustainably improve the livelihoods of many of the extremely poor
> rural people it targets. BRAC and a number of other organisations
> are currently replicating the programme's success in other countries,
> and early evidence of results is broadly positive.[54]

ICT applications – and specifically virtual payments – have made
social assistance programmes cheaper, safer and easier to implement,
and have extended their reach with less organisational infrastructure
than was required in the past. When I was involved in the implementation of a cash transfer programme for the UNHCR in Jordan in the
1990s, eligible people had to come to the Red Crescent Headquarters in
Amman, and queue to collect their monthly allowance. It was a cumbersome, time-consuming, disruptive, transaction-cost heavy and for
some people humiliating process. Nowadays, UNHCR-registered refugees in Jordan pay for their shopping by getting their iris scanned. This
is far less cumbersome (but comes with risks that I will return to in
Chapter 14).[55] Elsewhere, governments and organisations that provide
cash assistance provide eligible people with SIM cards, debit cards (with
or without biometric data incorporated into them), or transfers that are
linked to their digital identity and micro-ATMs (these are small hand-
held devices that have been adopted rapidly by rural shops). Once the
SIM or debit cards have been printed and distributed, transfers are done
at nearly zero cost. Incomings and outgoings are easy to monitor and
the money generally arrives without employees taking under-the-table
fees (though I came across a few accusations of such fees sometimes
being paid just to get on the beneficiary list). Robbers cannot steal from
cash depots because there *are* no cash depots.

In short, social assistance is a field in which ICT has helped
governments and other social assistance providers to streamline

processes, lower costs and reduce the risks of theft and corruption. ICT has had similar effects in other areas of what has come to be called 'e-government', which is covered in the next chapter.

Notes

1 For trends from 2011 to 2021, see GSMA (undated) *Mobile money metrics; global metrics*, GSMA, gsma.com/mobilemoneymetrics/#global, visited on 4 December 2022. (GSMA is a lobby group of some 800 mobile operators worldwide.)

2 Andersson-Manjang, S.K. and Naghavi, N. (2021) *State of the industry report on mobile money 2021*, GSMA, with absolute and 2020 growth figures by region on pages 8–9. In 2020, every indicator rose by between 5% and 38% in every region of the world, with only one exception – the number of transactions in the Middle East and North Africa – which declined in 2020.

3 For example, Seng, K. (June 2021) "The mobile money's poverty-reducing promise: evidence from Cambodia", *World Development Perspectives*, volume 22, article 100310. I give this particular example because this paper includes a literature review divided into literature that found positive and literature that found negative effects of mobile money on poverty.

4 Aron, J. (August 2018) "Mobile money and the economy: a review of the evidence", *The World Bank Research Observer*, volume 33, number 2, pages 135–188.

5 For a brief overview of international migration figures, see UN (17 September 2019) *The number of international migrants reaches 272 million, continuing an upward trend in all world regions, says UN*, United Nations. For the database underpinning this statement, including gender-disaggregated data, see bit.ly/Migration2019.

6 World Bank (2019) *Leveraging economic migration for development: a briefing for the World Bank Board*, World Bank, page 14 and Figure 2.6 on page 15.

7 According to the 2014 "Rural–urban migration monitoring survey" of the National Bureau of Statistics in China. See Démurger, S. and Wang, X. (2016) "Remittances and expenditure patterns of the left behinds in rural China", *China Economic Review*, volume 37, pages 177–190, figure on page 178.

8 Sharma, K. (2017) *India has 139 million internal migrants. They must not be forgotten*, World Economic Forum.

9 Generally, men dominate the migration pattern, but there are exceptions. The most notable exception is Sri Lanka where, in some years in the 1990s, there were three times as many women as men departing (a phenomenon sometimes referred to as the 'feminisation of migration'). For figures from 1991 until 2019, see slide 5 of ILO (February 2020) "Recent changes in labor migration trends and policies in Asia", *10th ADBI-OECD-ILO roundtable on labor migration in Asia*, International Labour Organization.

10 Apart from the pandemic's effects, there has only been one brief dip in a three-decades-long growth path in the global size of remittances. World Bank (2019) *Leveraging economic migration for development: a briefing for the World Bank Board*, World Bank, Figure 2.6 on page 15. This one brief

dip in remittances, which occurred in 2015 and 2016, caused short-lived worries – see World Bank (April 2017) *Remittances to developing countries decline for second consecutive year*, World Bank.

11 World Bank and KNOMAD (October 2020) "Phase II: Covid-19 crisis through a migration lens", *Migration and Development Brief 33*, World Bank and the Global Knowledge Partnership on Migration and Development, section 1.3 on pages 7–16.

12 UNGA (18 June 2018) *Resolution A/RES/72/281*, United Nations General Assembly.

13 Acosta, P. *et al* (2008) "What is the impact of international remittances on poverty and inequality in Latin America?", *World Development*, volume 36, number 1, pages 89–114, with the quotation on page 89. The conclusion in relation to inequality is not common, as migrants are not generally the poorest people in the village, and migration may well *increase* local inequalities. For an empirical study of this 'poverty trap' in Latin America see, for example, Golgher, A.B. (2012) "The selectivity of migration and poverty traps in rural Brazil", *Population Review*, volume 51, number 1, pages 9–27.

14 Imai, K.S. *et al* (2014) "Remittances, growth and poverty: new evidence from Asian countries", *Journal of Policy Modeling*, volume 36, pages 524–538.

15 Ellyne, M. and Mahlalela, N. (2017) *The impact of remittances on poverty in Africa: a cross-country empirical analysis*, paper presented at the 14th African Finance Journal Conference, 17–19 May 2017.

16 Sustainable Development Goals, target 10.c, indicator 10.c.1.

17 See the graph on page 2 of Pickens, M. (December 2009) "Window on the unbanked: mobile money in the Philippines", *CGAP Brief*, Consultative Group to Assist the Poor, 4 pages, emphasis added.

18 See Figure 1 on page 8 of Baffour, P.T., Abdul Rahaman, W. and Mohammed, I. (2020) "Impact of mobile money access on internal remittances, consumption expenditure and household welfare in Ghana", *Journal of Economic and Administrative Sciences*, volume 37, number 3, pages 337–354; Kikulwe, E.M., Fischer, E. and Qaim, M. (October 2014) "Mobile money, smallholder farmers, and household welfare in Kenya", *Plos One*, volume 9, issue 10, in the section on the conceptual framework, with references to research that dates back to 2009 – only two years after the launch of M-Pesa.

19 Hughes, N. and Lonie, S. (winter-spring 2007) "M-PESA: mobile money for the 'unbanked'; turning cellphones into 24-hour tellers in Kenya", *Innovations: Technology, Governance, Globalization*, volume 2, issue 1–2, pages 63–81, with both facts (the microfinance origins and the move towards a focus on remittances) mentioned on pages 77–78. This key proposition was defined during a workshop of the Safaricom commercial team, in 2006.

20 There are a few preliminary figures confirming this effect in the context of international rather than in-country remittances, as reported by Noha Emara and Yuanhao Zhang, on page 4 of Emara, N. and Zhang, Y. (May 2021) "The non-linear impact of digitization on remittances inflow: evidence from the BRICS", *Telecommunications Policy*, volume 45, number 4, 17 pages.

21 I am only aware of a single piece of research that attempts to isolate the effects of mobile money on migration decisions, in 102 villages in Mozambique, and it found that the introduction of mobile money into a rural area boosted people's propensity to migrate: Batista, C. and Vicentre, P.C. (February 2021) "Is mobile money changing rural Africa? Evidence from a field

experiment", *CReAM Discussion Paper Series 2116*, Centre for Research and Analysis of Migration. (This is an update of Working Paper 1805 of the Novafrica Working Paper Series, of December 2018.)

22 The finding that the switch from whatever previous system of transfers to mobile money transfers increases the frequency of transfers is common. Often, the finding is that mobile money causes the *chance* to receive remittances to increase, the *frequency* of these remittances to increase, *and* the *sum total* of the remittances to increase. See, for example, Munyegera, G.K. and Matsumoto, T. (2016) "Mobile money, remittances, and household welfare: panel evidence from rural Uganda", *World Development*, volume 79, pages 127–137.

23 Bastagli, F. *et al* (July 2016) Cash transfers: what does the evidence say? A rigorous review of programme impact and of the role of design and implementation features, Overseas Development Institute, page 32, with an indication of where to find more in footnote 30 at the bottom of that page.

24 This particular quotation is from Kikulwe, E.M., Fischer, E. and Qaim, M. (October 2014) "Mobile money, smallholder farmers, and household welfare in Kenya", *Plos One*, volume 9, issue 10, with the quotation from the section titled *descriptive statistics* (the paper does not have page numbering). Background: rainfall was low in Central and Eastern Kenya in 2009, and mobile transfers were significantly higher in that year than in the following year, when rain was more plentiful.

25 "Observation 4" on page 2 of Morawczynski, O. and Pickens, M. (August 2009) "Poor people using mobile financial services: observations on customer usage and impact from M-PESA", *CGAP Brief*, Consultative Group to Assist the Poor, 4 pages.

26 Patnam, M., Yao, W. and Haksar, V. (July 2020) *The real effects of mobile money: evidence from a large-scale fintech expansion*, International Monetary Fund, with the quotation on pages 5–6.

27 Most of the references used in this section on remittances report on this phenomenon, with the percentage of mobile phone owners using mobile money for this purpose ranging between 10% in the Philippines (in 2009) and 40% in Kenya (in 2014). For the first figure, see Pickens, M. (December 2009) "Window on the unbanked: mobile money in the Philippines", *CGAP Brief*, Consultative Group to Assist the Poor, 4 pages, page 4; for the second figure, see Kikulwe, E.M., Fischer, E. and Qaim, M. (October 2014) "Mobile money, smallholder farmers, and household welfare in Kenya", *Plos One*, volume 9, issue 10, Figure 2. In both cases the percentages will have increased by now, as m-wallet providers at the time had not yet thought of marketing their product as a saving tool.

28 See, e.g., Guermond, V. (December 2020) "Contesting the financialisation of remittances: repertoires of reluctance, refusal and dissent in Ghana and Senegal", *Environment and Planning A: Economy and Space*, 22 pages.

29 "Impact observation 1" on page 3 of Morawczynski, O. and Pickens, M. (August 2009) "Poor people using mobile financial services: observations on customer usage and impact from M-PESA", *CGAP Brief*, Consultative Group to Assist the Poor, 4 pages.

30 Baffour, P.T., Abdul Rahaman, W. and Mohammed, I. (2020) "Impact of mobile money access on internal remittances, consumption expenditure and household welfare in Ghana", *Journal of Economic and Administrative Sciences*, volume 37, number 3, pages 337–354.

31 Munyegera, G.K. and Matsumoto, T. (2016) "Mobile money, remittances, and household welfare: panel evidence from rural Uganda", *World Development*, volume 79, pages 127–137.

32 Farooq, S., Naghavi, N. and Scharwatt, C. (October 2016) *Driving a price revolution; mobile money in international remittances*, GSMA. The executive summary says that "In the 45 country corridors surveyed, the average cost of sending [US]$200 using mobile money was 2.7 percent, compared to 6.0 percent using global MTOs [Money Transfer Operators]". See also Kirui, B.K. (November 2022) "The role of mobile money in international remittances: evidence from Sub-Saharan Africa", *Research Paper 518,* African Economic Research Consortium, with a comparative graph in Figure 2.1 on page 6; and Rehman, F.U. and Nasir, M. (undated but 2020) *Measuring transaction costs of migrant remittances in Pakistan*, Federal SDG Support Unit, which finds that "the introduction of mobile money as an instrument of transaction has decreased the previously high cost of remittances" (quotation from page 42).

33 For example, over 70% of formal remittances between New Zealand and Tonga are processed through mobile money service. See Granryd, M. (20 September 2018) *More than just a phone: mobile's impact on sustainable development*, World Economic Forum. A qualitative piece of research that is based on interviews with 10 Zimbabweans who work illegally in South Africa suggests that sending remittances by mobile phone was standard practice and that "the requirements for opening a *mukuru.com* account was not cumbersome as any valid identification document such as [a] Zimbabwean passport and valid phone number could be used". Mavodza, E. (2019) "Mobile money and the human economy: towards sustainable livelihoods for Zimbabwean migrants in South Africa", *Africa Development*, volume XLIV, number 3, pages 107–130, with the quotation from page 114.

34 Ratha, D. *et al* (October 2019) "Data release: remittances to low- and middle-income countries on track to reach $551 billion in 2019 and $597 billion by 2021", *World Bank Blogs*, World Bank, with the quotation in the last paragraph of the blog.

35 The volume of international payments by mobile phone does grow fast: from US$3.1 in 2015 to US$12.7 billion in 2020. Andersson-Manjang, S.K. and Naghavi, N. (2021) *State of the industry report on mobile money 2021*, GSMA, with the figures from Figure 15 on page 37.

36 FATF (2016) *Guidance for a risk-based approach for money or value transfer services*, Financial Action Task Force, Paris. FATF is an inter-governmental body that sets standards in relation to measures to combat money laundering, terrorist financing and other threats related to the international financial system.

37 Oyoo, G.O. "Impact of mobile money on financial crime, money laundering, and terrorism financing", Chapter 10 of Opati, T.Z. and Gachukia, M. K. (2020) *Impact of mobile payment applications and transfers on business*, IGI Global, pages 213–230, with the quotation from page 226.

38 Bureau for International Narcotics and Law Enforcement Affairs (March 2014) *International narcotics control strategy report; volume II, money laundering and financial crimes*, United States Department of State. The statement that comes closest to Oyoo's assertion is this one, from page 184: "Al-Shabaab moves some of its funds via cash couriers, but a significant

portion reportedly passes through hawaladars and other money or value transfer services".

39 *Ibid*, with the quotation on page 125.

40 Tina Rahimy, a dear friend who commented on an early version of this book, wrote here that "when I open a bank account in the Netherlands I have to solemnly promise not to send money to Iran – so to my *family* – or the bank will not allow me to open an account".

41 For preliminary insights in the way blockchain technology could provide a step change in further cost reductions, see Rühmann, F. *et al* (April 2020) "Can blockchain technology reduce the cost of remittances?", *OECD Development Cooperation Working Paper 73*, Organisation for Economic Co-operation and Development. This paper also highlights a number of blockchain-related risks.

42 See, for example, vodacom.co.tz/international; and its terms and conditions.

43 See, e.g., Farooq, S., Naghavi, N. and Scharwatt, C. (October 2016) *Driving a price revolution; mobile money in international remittances*, GSMA.

44 For example, the GSMA Mobile Money Programme is supported by the Bill & Melinda Gates Foundation, the MasterCard Foundation and the Omidyar Network (which is both a company and a foundation that gets its funding from the Pierre Omidyar Trust).

45 Farooq, S., Naghavi, N. and Scharwatt, C. (October 2016) *Driving a price revolution; mobile money in international remittances*, GSMA. In their executive summary, they report that "global MTOs tend to offer their services at lower prices in markets where they are in competition with mobile money providers (6.0 percent compared to 8.2 percent)". The same claim is made for fintech companies in general (so including other newish forms of electronic transfers of remittances): see Cortina, J.J. and Schmukler, S.L. (April 2018) "The fintech revolution: a threat to global banking?", *Research and Policy Briefs number 14*, World Bank. On page 3, they report that "following the increasing use of fintech providers, the cost of sending remittances has been declining [...] while the speed of transactions has been increasing".

46 On 20 November 2022, the GSMA website (gsma.com) gave 4,550 hits on the search term "SDG".

47 ICAI (April 2019) *DFID's partnerships with civil society organisations; a performance review*, Independent Commission for Aid Impact; see paragraph 4.70. In this book I use the terms CSOs and NGOs interchangeably.

48 See also Hayes, B. (2017) *The impact of international counter-terrorism on civil society organisations: understanding the role of the Financial Action Task Force*, Bread for the World.

49 FATF (October 2001) *IX special recommendations*, Financial Action Task Force.

50 FATF (updated version of October 2021) *International standards on combating money laundering and the financing of terrorism & proliferation: the FATF recommendations*, Financial Action Task Force, with the quotation on page 13.

51 World Bank (2018) *The state of social safety nets 2018*, World Bank, see foreword, page 16. This is the most recent issue at the time of finalising this book (September 2023). This average hides major differences across countries. I do not know of databases that express social assistance expenditures as a percentage of government expenditure.

52 ICAI (January 2017) *The effects of DFID cash transfer programmes on poverty and vulnerability; an impact review*, Independent Commission for Aid Impact, with the quotation on page 33.

53 *Ibid*, with the quotation from paragraph 4.17 on page 16.

54 Banerjee, A. *et al* (15 May 2015) "A multifaceted program causes lasting progress for the very poor: evidence from six countries", *Science*, volume 348, issue 6236, page 772.

55 UNHCR started using iris scans in Jordan in 2012 (ten years after its first pilot, which took place in Pakistan in 2002) and UNHCR's Heba Azazieh hailed the innovation as a success as "now we have 100 percent certainty that the money reaches those it was intended for". Quoted in UNHCR Innovation (30 August 2016) *Using biometrics to bring assistance to refugees in Jordan*, United Nations Refugee Agency. In 2012, refugees needed to use a debit card in combination with an iris scan, but at the time of writing this book the EyePay technology UNHCR uses in Jordan no longer requires debit cards.

6 E-government

Ever since the 'dot-com era' of the 1990s, governments in the Global South have been investing heavily in e-government. Definitions vary, but the World Bank's description of e-government is often used:

> E-Government refers to the use by government agencies of information technologies [...] that have the ability to transform relations with citizens, businesses, and other arms of government. These technologies can serve a variety of different ends: better delivery of government services to citizens, improved interactions with business and industry, citizen empowerment through access to information, or more efficient government management.[1]

Because of the rural–urban digital divide (i.e., the gulf between those who easily access and use modern ICT and those who do not – see Chapter 12), awareness and usage of the citizen-focused side of e-government is likely to start with well-connected and educated urbanites. However, rural people could potentially benefit more than these urbanites, as they are further away from urban centres and could therefore potentially save more time by arranging their affairs in a local online shop or telecentre instead of having to travel to a city-based government building.

This is only true if e-government services actually work, which is not always the case. In the early stages, e-government initiatives were a matter of trial and error. Many initiatives collapsed, or failed to reach the poorer segments of communities – so much so that, in 2003, Richard Heeks published a paper titled "Most eGovernment-for-Development projects fail: how can risks be reduced?".[2] In this paper, Heeks assessed an unspecified (but "over 40") number of reports on e-government cases from "developing and transitional countries" and concluded that most e-government projects failed. He posited that such failures were often

DOI: 10.4324/9781003451716-8

caused by incorrect assumptions. For example, if an e-government design assumes that the availability of information improves government decision-making while, in reality, decision-makers use their gut feeling instead, then an e-government investment in information provision to government decision-makers is likely to fail.[3]

Notwithstanding the frequent initial failures, governments continued to invest in e-government, and their persistence was encouraged by the World Bank, which pushed the mantra that "resulting benefits can be less corruption, increased transparency, greater convenience, revenue growth, and/or cost reductions".[4] Lessons were learned and, by now, many countries' e-government is helping people by reducing red tape, ambiguity and opportunities for bribery.

Two examples from my personal experience seem worth sharing. They are from Egypt and India. I lived in both countries in the early 1990s, prior to the launch of their e-government systems, and visited both countries regularly after their e-government systems were introduced and had time to mature. Both countries had a low ranking on the *Bureaucratic Efficiency Index* in the time before e-government,[5] and both governments have since introduced a range of online products and services.

- When I lived in Egypt for the first time, in 1994, arranging anything that involved the government was difficult. I spent many hours waiting, getting shrugs and incorrect information from employees, filling forms and getting them stamped in offices across Cairo. The Mugamma, a building in the centre of Cairo, was the epicentre of the country's bureaucracy and the only government building where I truly lost my cool and yelled at people, without effect of any kind. From remote rural regions, a visit to the Mugamma could amount to a two-day journey each way. Its reputation of opacity and senseless queueing was such that it provided the plot for a 1992 movie in which a man, rattled by his fruitless queueing, accidentally takes the building hostage (*Terrorism and Kebab*, sometimes referred to as the most popular Egyptian movie to date[6]). Nowadays, the most common services – renewing a vehicle licence, paying bills, fees and fines, getting train tickets, checking library catalogues, getting registered to get married – are available online. The Mugamma is no longer operational and visits to other government buildings are far less commonly needed, and easier than they used to be. It is now possible to pick up a birth certificate from anywhere in the country, for example, rather than only from the city hall where the birth was originally registered; and the preparatory work (form filling, fee payments) can be done online and prior to the visit.

- When I spent six months in India, in 1992–93, India's corruption was well-researched[7] and such an everyday reality[8] that bribes were paid even where they were unnecessary, because of the *assumption* they would be needed (see Box 6.1). Bureaucracy was impenetrable without the help of intermediaries and, worse, red tape and these intermediaries had the tendency to reinforce each other, as red tape provided opportunities for intermediaries, and the presence of intermediaries incentivised bureaucrats to add further red tape.[9] I got a taste of this impenetrable bureaucracy when I bought a second-hand auto rickshaw and tried to get an auto rickshaw driving licence. After six weeks of queueing, form-filling and paying small sums here and there (never quite sure if they were bribes or fees) I had come no closer to a licence, and had not managed to gain insight into the process, timelines, exams and costs involved. Marianne Bertrand and her colleagues showed, in a piece of 2007 research, that I should have engaged a private intermediary (an 'agent'), as agents know who and how to bribe, sometimes to *accelerate* but also to *circumvent* the formal process. In their sample, only one in eight people who obtained their licence with the help of an agent ever took an actual driving test, against 94% of those who did not use an agent.[10] After India's investments in e-government the process towards a driving licence became far clearer, and nowadays it is outlined and largely conducted online.[11] When driving tests in Bangalore, in particular, were shown to be highly corruption-prone, the Department of Transport used another ICT solution to eliminate that, too, by conducting the entire driving test as a simulation.[12]

Box 6.1 Bribes in the expectation of corruption

The regions of India that had undergone land consolidation processes were the first to benefit from the Green Revolution. The benefits of consolidation were so significant and obvious that farmers, who had first resisted it, began to demand it a few years into the process.

In Uttar Pradesh, the consolidation process was a very open one and had multiple stages of appeal. The design of the process, as well as a few checks and balances within the government department responsible, meant that the "features of the programme reduce[d] both the influence of the large and powerful farmers of the village and the significance of corruption". Nonetheless, farmers paid large sums to various types of intermediaries, who mostly just pocketed the money, simply because the farmers fully *expected* the process to be corrupt and bound to lead to adverse results unless bribes were paid.

Source: the text in this Box is based on Oldenburg, P. (July 1987) "Middlemen in Third World corruption: implications of an Indian case", *World Politics*, volume 39, issue 4, pages 508–535, with the quotation from page 517.

The progress of e-government in Egypt and India was underpinned by a national roll-out of digital IDs. By 2018, the two countries had provided digital IDs to 90% and 98% of their populations, respectively. They are not the only countries in the Global South that have rapidly increased their digital identification coverage. China, Thailand, Indonesia, Brazil and several other large countries in Asia and Latin America are in the same ballpark, and the trend suggests that many other countries are following suit.[13] While the percentages do not differentiate between urban and rural people, they are so high by now, in each of the countries just mentioned, that the vast majority of rural people must have a digital ID even if there is an extreme rural–urban digital divide (i.e., if *all* the people without digital ID were to be rural people, rural coverage would still be well above 90%).[14]

The ways countries utilise digital identification go well beyond the Egyptian and Indian examples I just gave. Where e-government services are well-organised, digital identification may help people to get what they need from the public authorities in a relatively swift, unbureaucratic, unambiguous, affordable and corruption-proof manner. E-government facilitates the roll-out of social assistance programmes, including both unconditional cash transfers and people's enrolment in rural job guarantee schemes. The digital ID itself may merely be a number, in combination with a fingerprint, iris scan and basic personal data, but it can potentially be linked with almost any other type of data. Where this happens (as is the case in India and China, for example), it also helps with non-governmental parts of life, through know-your-customer systems that use personal data linked to people's digital identity. The Indian know-your-customer system (e-KYC),[15] for example, eased the due diligence requirements for banks and other businesses, and drove down the costs of client onboarding from the equivalent of US\$5 to US\$0.70;[16] and e-Sign functionality enables people to sign documents digitally – to name but two of an increasing number of facilities that save time, money and hassle.

Digital identification is affordable (in India the costs are around US $1.50 per registered person[17]) and is useful for governments as well, as it helps them to reach their citizens more easily. This can be for things many citizens welcome, like social assistance services, useful announcements or warnings, or ballot papers. It can also be for things that citizens may not particularly welcome but that are useful nonetheless. Digital identification helps with the formalisation of the economy, enables electronic road toll payments, and helps to ensure that people are aware of and pay the fines imposed on them. E-government may also facilitate a country's citizens' taxation process, especially in countries where ICT adoption by citizens is relatively high[18] – though, as the Dutch childcare benefits scandal illustrates,[19] e-government will not resolve all the many tax-related challenges governments face (see also Box 6.2). Such challenges are likely to have the most dramatic effects in countries without critical media and with the least well-resourced and developed institutions.

The high rates of digital IDs in rural regions show that, when it comes to digitalisation of identities, rural regions do not need to lag behind towns and cities. India, for example, made a concerted and successful effort to digitalise the identity of its rural citizens – the first person to receive an 'Aadhaar' number (i.e., the 12-digit number that represents a person's digital identity) was not a VIP from New Delhi but a woman from the village of Tembhli, in Nandurbar District, Maharashtra.[20] Provided there is electricity and 3G coverage (and for some government services even 2G coverage suffices), rural people stand to benefit even more from virtual services than their urban counterparts. This is because they otherwise have to travel further to arrange their affairs related to land registration, inheritance and ownership transfer, or the registration of births, deaths, marriages or businesses. These benefits still often fail to reach the poorest segments of rural society, as these remain underrepresented among users of e-government facilities. However, their access is improving, in part because companies and small shops have started catering for their e-government needs. For example, I recently assessed and was impressed by a rapidly growing company – OneBridge – that brings e-government services and online shopping together on a single intermediary app that requires only basic literacy. The company provides this app to rural shop owners in parts of India, and offline individuals go there to, for example, pay their electricity bills. Such roll-out presents significant time savings: Olivia White and her colleagues estimate that they "could save citizens around the world an average of 20 hours per year", and that the reduced "need to travel [is] a particular benefit for people who live in rural areas".[21]

In addition to the reduction in red tape, there is evidence that the net effect of e-government has been a reduction in *corruption* within citizen-focused bureaucratic processes[22] (but see Box 10.1). This is the case particularly where there is a minimum level of institutional quality and 'digital literacy' within the country.[23] Where a corruption-reducing effect exists, it may benefit rural people more than urbanites, as their lower average levels of education and their greater distance to bureaucratic centres may mean their vulnerability to corruption is particularly high – but this is merely a hunch because, to my knowledge, no comparative research has been conducted in this field.

Box 6.2 Airtel Money facilitates tax collection, but itself evades taxation

In October 2020, the Malawi Revenue Authority (MRA or 'the Authority') announced a partnership with Airtel Money, to enable taxpayers to arrange and pay their domestic taxes and custom duties via their phone. In a statement, MRA explained that it "is leaving no stone unturned to enhance tax compliance by opening up more avenues for taxpayers to timely manage their obligation of paying tax. At the same time, the Authority is focussed on reducing the cost of tax compliance while improving revenue collection to avail adequate resources to Government for socio-economic development".

A few months later, in February 2021, the Centre for Investigative Journalism Malawi published a paper that outlined the many ways in which Airtel Money, through its parent company, had itself avoided paying millions of dollars in taxes in Malawi – in large part through complicated loans and other transactions across the parent company's many subsidiaries, often involving a Dutch holding company.

Source: the text in this Box is based on MRA (8 October 2020) *MRA partners Airtel Money for tax payments*, Malawi Revenue Authority, with the quotation in paragraph 4; and CIJM (8 February 2021) *Airtel Malawi books raise red flags over international tax evasion*, Centre for Investigative Journalism Malawi. In this Box I cover *Airtel Money* because of the paradox of a company that works with a national tax office whilst itself evading taxation – but it is only one of many companies that abuse loopholes in tax systems to evade taxation in African countries. SportPesa, a betting company discussed in Chapter 11, is also one of these companies – see Foul, D. *et al* (30 March 2021) "SportPesa profiteert van gokmanie in Kenia", *Trouw*, supplement titled "Money Trail", pages 6–8 [SportPesa benefits from betting mania in Kenya].

ICT could also help reduce the scope for corruption in other ways than through standardised e-government processes. First, governments could make information – related to people's rights and entitlements and public procurement, for example – publicly available. Such practice may in and by itself reduce the scope for corrupt behaviour, just as the non-ICT-related openness described in Box 6.1 did in relation to land consolidation in Uttar Pradesh (where openness took the shape of open meetings). In addition, public information allows social accountability mechanisms to monitor compliance. This is because information portals that tell people how much money has been allocated to, say, a village's electrification contract, and what conditions must now be fulfilled, may increase the likelihood that the village will indeed get hooked up to the electricity network. Second, individuals are able to give visibility to corruption via online platforms such as 'I paid a bribe', a platform that started in India, in 2010 and that now exists in a range of countries. (Other such crowdsourcing platforms exist as well – in Vietnam there is Toidihoilo, for example.) These platforms started in part because the traditional mechanisms of direct complaints via hotlines, one of the earlier ICT-driven accountability mechanisms, only rarely worked well (see Box 6.3).

Box 6.3 Hotlines often do not work

Governmental and non-governmental service providers often have some sort of hotline that users (or 'beneficiaries', in outdated but still often-used terminology) can call to report corruption, sexual abuse and other forms of misconduct. This is the longest-standing ICT accountability mechanism that I am aware of and, if my evaluations are collectively representative, it rarely works. The hotline number's existence and use are two things I always check when evaluating mobile health clinics or social assistance programmes, for example. Conversations about this generally start off promising and end up disappointing: "Yes, absolutely, we have a dedicated phone line for this." "Great! What is the process for handling these complaints, and could I see last month's records please?" My respondents then nearly always admit that the phone line is not currently operational, or at least that phone calls are not logged and systematically followed up on.

The popular press tends to be enthusiastic about such ICT-driven social accountability mechanisms,[24] and so does the World Bank, a long-standing champion of investments in good governance. In the 2021 issue of its flagship *World development report*, the World Bank gives only two concrete examples of success, and neither stands up to scrutiny. First, the report celebrates the achievements of the World Bank-supported Africa Freedom of Information Center,[25] but it does so with reference to a 2018 report that suggests a positive verdict is premature because, four years after the start of the project, "only 5.8% of the total planned contracts were awarded and of all the contracts that were monitored, none started on their set start date".[26] Second, the report implies the aforementioned 'I paid a bribe' platform is successful, as it says that "it empowers individuals, civil society, and governments to fight corrupt behavior" and that "local leaders can act against corrupt officials and receive real-time feedback on the impact of anticorruption policies".[27] However, independent research concludes that "the presence of a crowdsourced corruption reporting platform like the 'I paid a bribe' website does not in fact lead to systematically lower bribery as compared to a setting where no bottom-up reporting is allowed".[28] A review of evidence of the corruption-reducing effects of ICT, conducted by an organisation with less bias, came to a less favourable verdict than the World Bank's flagship publication, and added the qualifier that far more research in this field is needed before firm conclusions can be drawn.[29]

Notes

1 World Bank (May 2015) *E-government brief*, World Bank, and repeated many times since.
2 Heeks, R. (2003) "Most eGovernment-for-Development projects fail: how can risks be reduced?", *iGovernment Working Paper Series*, number 14, Institute for Development Policy and Management.
3 Heek's seven dimensions are (1) information, (2) technology, (3) processes, (4) objectives and values, (5) staffing and skills, (6) management systems and structures, and (7) "other resources". Heeks called them the ITPOSMO dimensions.
4 This quotation originates from page 5 of Fang, Z. (2002) "E-government in digital era: concept, practice and development", *International Journal of the Computer, the Internet and Management*, volume 10, issue 2. It has been quoted in a number of World Bank and other publications, such as World Bank (May 2015) *E-government brief*, World Bank; and Dada, D. (2006) "The failure of e-governance in developing countries: a literature review", *the Electronic Journal on Information Systems in Developing Countries*, volume 26, number 7, pages 1–10.
5 The Bureaucratic Efficiency Index is the average of three Business International indices, covering the judiciary system, red tape and corruption. See Table 1 on page 687 of Mauro, P. (August 1995) "Corruption and growth",

The Quarterly Journal of Economics, volume 110, number 3, pages 681–712. In an index that scores a maximum of 10, Egypt was in the 1.5–4.5 bracket (the lowest category, of six), and India in the 4.5–5.5 bracket (the second lowest bracket).

6 Sardar, Z. (2 April 2007) *What Egyptian cinema can teach us*, New Statesman.

7 See, for example, Palmier, L. (1985) *The control of bureaucratic corruption: case studies in Asia* [India, Hong Kong and Indonesia], Allied Publishers; and, for a thorough description of truly systematic corruption that covers multiple administrative layers, Wade, R. (April 1982) "The system of administrative and political corruption: canal irrigation in South India", *Journal of Development Studies*, volume 18, number 3, pages 287–328. Although the latter paper's research is specific to a single (unnamed) state in southern India, the paper also points out that "several bits of evidence suggest that the practices described here are not confined to our particular state" (quotation from page 317).

8 Philip Oldenburg gives an apt illustration of just how common bribing is, in the form of a quotation from the astrology column of the Hindustan Times of 13 October 1985: "Virgo: [...] If paying a bribe to anyone, see that the job is done". Oldenburg, P. (July 1987) "Middlemen in Third World corruption: implications of an Indian case", *World Politics*, volume 39, issue 4, pages 508–535, with the quotation from the top of the first page.

9 Fredriksson, A. (May 2014) "Bureaucracy intermediaries, corruption and red tape", *Journal of Development Economics*, volume 108, pages 256–273.

10 Bertrand, M. *et al* (November 2007) "Obtaining a driving license in India: an experimental approach to studying corruption", *The Quarterly Journal of Economics*, volume 122, issue 4, pages 1639–1676, with the percentage in Table V, on page 1662. The solution my friend and I came up with at the time was that we typed up a few words indicating that we were the owners of the rickshaw with number plate so-and-so, and got it laminated. This got us through the various checkpoints and spot checks.

11 See transport.delhi.gov.in/content/online-appointment-and-payment-driving-license.

12 One World Foundation India (September 2011) *Automatic driving test track*, oneworld.net.

13 White, O. *et al* (April 2019) *Digital identification: a key to inclusive growth*, McKinsey Global Institute, Exhibit E1 on page 3 and Exhibit 4 on page 27. Note that "Exhibit E1" is a somewhat confusing infograph. The dark blue font for "ID but no digital trail" is about social media use, not about IDs that exist in hard copy format only; and the 90% for India does not come from the charts but from *Note 5* underneath the chart. Note also that the report incorrectly assumes that people who are not active on social media do not use their digital ID. In my experience, many rural people do not use internet themselves but do go to online rural shops, where they pay the shopkeeper to use e-government and mobile money services on their behalf. Lastly, for India this 'exhibit' conflicts with another source: Misra, P. (January 2019) "Lessons from Aadhaar: analog aspects of digital governance shouldn't be overlooked", *Background Paper Series number 19*, Pathways for Prosperity Commission, which says, on page 3, that 99% of the (adult) population has a digital ID.

14 For example, in 2020 less than 35% of the people in India were urban, so even in the most extreme case where 100% of urban people had a digital ID, 97% of rural people would have a digital ID as well, to come to the estimated 98% overall digital ID ownership. The figure on the proportion of people in India living in urban regions is from Statistica (June 2021) *India: degree of urbanization from 2010 to 2020*, Statistica.

15 In India this works through an API (Application Programming Interface) called IndiaStack – see indiastack.org/about.

16 This is reported in Gelb, A. and Diofasi Metz, A. (2018) *Identification revolution: can digital ID be harnessed for development?*, Centre for Global Development, on page 175.

17 Nilekani, N. and Shah, V. (2015) *Rebooting India; realizing a billion aspirations*, Penguin Books India, with the fact mentioned and explained on page 36.

18 Uyar, A. *et al* (2021) "Can e-government initiatives alleviate tax evasion? The moderation effect of ICT", *Technological Forecasting & Social Change*, volume 166, 13 pages.

19 The 'Toeslagenaffaire'. For a concise overview, see Henley, J. (14 January 2021) "Dutch government faces collapse over child benefits scandal", *The Guardian*. For the full report, see Tweede Kamer der Staten-Generaal (December 2020) "Ongekend onrecht; verslag Parlementaire Commissie Kinderopvangtoeslag", *Kamerstuk 35 510*, number 3, Tweede Kamer der Staten-Generaal [Unprecedented injustice: report of the Parliamentary Commission on Child Care Support].

20 For this fact and a description of the Aadhaar roll-out, see chapter 1 of Nilekani, N. and Shah, V. (2015) *Rebooting India; realizing a billion aspirations*, Penguin Books India.

21 White, O. *et al* (April 2019) *Digital identification: a key to inclusive growth*, McKinsey Global Institute, with the (unsubstantiated) quotations from pages 93 and 43 respectively.

22 "The literature on e-government [...] finds a clear statistical relationship between different measures of e-government adoption and reduced corruption." Adam, I.O., and Fazekas, M. (December 2018) "Are emerging technologies helping win the fight against corruption in developing countries?", *Pathways for Prosperity Commission Background Paper Series,* number 21, with the quotation from page 2.

23 Adam, I.O. (May 2020) "Examining e-government development effects on corruption in Africa: the mediating effects of ICT development and institutional quality", *Technology in Society*, volume 61, 10 pages. Digital literacy is "the ability to access, manage, understand, integrate, communicate, evaluate and create information safely and appropriately through digital technologies for employment, decent jobs and entrepreneurship. It includes competences that are variously referred to as computer literacy, ICT literacy, information literacy and media literacy". See page 6 of Law, N. *et al* (June 2018) "A global framework of reference on digital literacy skills for Indicator 4.4.2", *Information Paper*, Number 51, United Nations Educational, Scientific and Cultural Organization (UNESCO).

24 Such as in Strom, S. (6 May 2012) *Websites shine light on petty bribery worldwide*, The New York Times; and Crawford, C. (19 May 2015) *Crowdsourcing anti-corruption: 'Like Yelp, but for bad governments'*, The Guardian.

25 World Bank (2021) *World development report: data for better lives*, A World Bank Group Flagship Report, page 58.

26 AFIC (April 2018) "Eyes on the contract: citizens' voice in improving the performance of public contracts in Uganda", *2nd Monitoring Report*, Africa Freedom of Information Center.

27 World Bank (2021) *World development report: data for better lives*, A World Bank Group Flagship Report, page 10 and 130 respectively.

28 Ryvkin, D., Serra, D. and Tremewan, J. (2017) *"I paid a bribe*: an experiment on information sharing and extortionary corruption", *European Economic Review*, volume 94, pages 1–22, with the quotation from page 14. Based on laboratory research, the researchers do offer advice on ways in which such platforms *could* potentially reduce extortionary bribes. Essentially, the paper suggests including specific information about the office that – or employees who – asked for a bribe, which would probably have a deterrent effect but would also be highly vulnerable to abuse, which is why 'I paid a bribe' does not have this feature.

29 Adam, I.O., and Fazekas, M. (December 2018) "Are emerging technologies helping win the fight against corruption in developing countries?", *Pathways for Prosperity Commission Background Paper Series*, number 21.

7 Climate change and natural resource management

There are ample ICT applications that monitor climate change. Among much else, they follow the build-up and direction of storms and rain patterns; monitor the thickness of the ozone layer; measure surface, air and ocean temperatures; and provide the processing power needed for the generation of multi-model projections for each of these phenomena. We know climate change is a reality, and things are getting worse. We also know that climate change amplifies threats to rural livelihoods. Among other things, it adds volatility to the weather, drives up the average temperature, prolongs droughts and facilitates the formation of hurricanes and tornadoes.

Techno-optimists believe that this is about to change, as ICT applications will help to decelerate and possibly reverse climate change. In principle, there are plenty of opportunities for this. For example:

- Applications that intensify farming could potentially reduce the need for farmland, and if this is combined with ambitious renaturing then the planet would regain some of its lost carbon storage capacity.
- Harvest storage and trading applications that reduce wastage and enhance logistical efficiency could potentially reduce greenhouse gas emissions.
- ICT-powered lockout technology could turn the supply of off-grid electricity into a commercially evermore viable option, and this could reduce the need for generators powered by fossil fuels.
- Mobile money facilitates the transfer of remittances and thereby potentially reduces the need for trips to the village.

However, so far none of this has led to a deceleration of climate change. First, some behavioural change is inconvenient or costs money, and this is a problem because "protecting the environment is [...] supported by a

DOI: 10.4324/9781003451716-9

Climate change amplifies threats

large majority – it's just not supported *very strongly*".[1] Second, energy efficiency gains only have a potentially positive effect on the climate change trajectory if production levels remain constant. This is not currently the case, which means that the net result of such ICT applications may well be an increase in overall emissions as the relative decoupling is more than compensated by the rebound effect (see Box 7.1).

Box 7.1 Do ICT-facilitated efficiency gains reduce the pressure on natural resources?

The hope that is expressed in the *Borlaug Hypothesis* is that efficiency gains reduce the use of resources. "The idea [is] that increasing crop yields can help prevent cropland expansion and deforestation, thus alleviating hunger and poverty without dramatically increasing environmental impact".[2] In reality however, yield gains have not generally followed the Borlaug Hypothesis at all. Instead yield gains tend to cause a pattern that is called the *Jevons paradox*, which was first observed in 1865:[3] efficiency gains in the use of a resource lead to more, not less, consumption of that resource. In other words, the fixed agricultural production level assumption is not realistic in a market-driven system that assigns no value to unused lands. Palm oil is the embodiment of the Jevons paradox (nowadays better known as the *rebound effect*): after achieving efficiency gains in the production of oil palm and the processing of palm oil, cultivation rapidly expanded and oil palm came to cover 10% of the world's crop land. The crop is responsible for 8% of the world's deforestation caused by the expansion of agricultural lands.[4]

One of the adverse effects of climate change is that it adds pressure onto the world's 'common resources'. Common resources are fishing grounds, forests, grazing lands and water sources that are (1) not in private ownership or owned and actively managed by government; (2) renewable, maintainable or fixable if they are managed wisely; but often (3) subtractable in the short term, which means that the same good cannot be used twice (e.g., if you pick a mushroom, that mushroom is gone). This subtractability poses a problem because common resources are also (4) non-excludable, which means that it is difficult to prevent people accessing them, should they choose to do so. Many of these common resources are deteriorating because humankind is over-using them... and climate change accelerates that deterioration. Wildfires are not generally caused by climate change, for example, but climate change causes drought and heat and this may make it harder to terminate them. The world's fish population is in decline, not because of climate change but because of overfishing, but climate change adds to the pressure as, say, lake fish may die off as the lake's temperature increases.

There are many ICT applications that monitor the state of the planet's common resources. Satellite imagery helps monitor the size and health of forests and savannahs, the size of lakes and the state of coral reefs. Drones and cameras add detail. Soil sensors transmit data about the soil's moisture, oxygen levels and chemical composition (though obviously only in the precise location of the sensor, which illustrates that a high volume of data does not necessarily mean representativeness of data). There are ICT applications that measure air pollution (where the location matters less than in the case of soil sensors, because of wind), track fish migration flows (and even, through tagging, individual animals) and do various other things that help provide insight into the state of and changes in the natural world and the space surrounding it. Some applications are used to monitor the results achieved by specific natural resource management systems as well. For example, researchers in Ghana used time series of satellite images to confirm that the community-based natural resource management system (or CBNRM system – see Box 7.2) covering the Afadjato-Agumatsa Community Nature Reserve had successfully conserved its forests.[5]

Box 7.2 Community-based natural resource management system

The premise of 'community-based natural resource management systems', or 'CBNRM systems', is that the management of resources that cover large territories, such as forests, savannahs and fisheries, is impossible without the participation of the people living on and around these resources. Moreover, for their participation to be useful such systems need to be broadly aligned with local values, customs and sense of what justice looks like. Where this is the case, CBNRM systems may serve three objectives. First, they may improve the livelihoods of those who use these resources, by ensuring reasonable access to all users. Second, they may ensure that these resources remain available, by avoiding usage beyond the resources' immediate renewability. And third, such systems may reduce conflict over these resources because people adhere to what they perceive to be credible rules.

There are many CBNRM systems, throughout the Global South. Some of them work reasonably well, but failures are also common. The overall spirit of high hopes, for CBNRM systems in particular[6] and for participatory approaches in general,[7] is long gone, but a better system of natural resource management has yet to announce itself.[8]

Participants in CBNRM systems themselves may also use ICT applications. In Tanzania and elsewhere, members of CBNRM systems that participated in REDD+ pilots captured their field observations using OpenDataKit[9] (ODK) – an often-used open-source data collection software that works offline and syncs when a connection is found. More commonly, text messages and radio shows raise the visibility of rules and sanctions; people who verify compliance to community agreements use photos and clips to evidence their findings; and websites and booking apps help CBNRM systems to generate money through tourism. There is a (no longer active) CBNRM group that has a modest presence on a few social media platforms (e.g., linkedin.com/company/cbnrmnet), and for a while (2000–08) there was an online Community of Practice on CBNRM (cbnrm.net).

The use of such ICT applications in common resource management is rarely reported in the literature. This is because it is nothing special: none of these applications are specific to common resource management, and all of the 30 tools recommended by the Community of Practice website[10] are used in a wide range of other contexts as well.

In principle, much more specialist ICT applications could be adapted for CBNRM systems. Vessel Monitoring Systems (VMS) are often used to track commercial fishing vessels, but could equally be used by CBNRM systems to track small local fishing boats. Satellite imagery that monitors the greenness of farmland (for the purpose of index insurance or to predict harvest prices) could also be used to support CBNRM decisions about sustainable foraging volumes. A nameless prototype tool that translates complex data in a user-friendly app that aims to inform moose-hunting management in Sweden[11] could be modified to support decisions made by communities in relation to, say, the hunting of wildebeest, a species with similarly extended roaming grounds as moose. Some of the many and sometimes long-standing volunteer-based biological recording schemes[12] could also be adapted to inform hunting decisions. Apps such as Picture This, which I often use during hikes to identify unfamiliar plant species, could be adapted to support foragers.

Optimists see opportunities, but to my knowledge there have been few if any actual cases where an ICT solution was adapted to suit a CBNRM system, and which the CBNRM system then adopted and actually used. Mangana Rampheri and Timothy Dube make a case for CBNRM systems using geographic information systems (satellite imagery and large-scale statistical operations that jointly provide insight into, say, the health trend of forests) as a basis for decisions,[13] but I do not know of cases where this has actually happened. REDD+

programmes experimented with remote sensing technology to support community-based forest management systems, but it did not work well.[14] Attempts to introduce biological monitoring apps among rural populations in the rural Global South have also, by and large, been unsuccessful.[15]

Moreover, applications that could potentially be useful are more likely to be used to *harm* than to *conserve* common resources. Illegal hunters may be more likely to use drones for hunting purposes than CBNRM systems are likely to use them to prevent illegal hunting, for example, and the fishing industry is more likely to use echo sounders to find and *catch* fish than CBNRM systems are likely to use them to track and *protect* fish. This is because illegal activities typically require the involvement of fewer people than CBNRM systems, so trust issues are easier to manage, and for the individuals involved the direct rewards of illegal hunting and fishing are likely to be higher than the direct rewards of a collective management effort.

The reason that CBNRM-specific ICT adaptations do not exist and that CBNRM systems use nothing but the most basic ICT applications is that rural communities that depend on common resources are at the wrong end of a multi-layered digital divide. Forest and fishery communities often have relatively little formal education.[16] They are often located in such remote regions that connectivity is poor even compared to other rural settings, and they are not reached by 'community information centres' (CICs) or internet kiosks.[17] Dispersed forest populations often speak minority languages that apps, text message services and radio shows do not normally use and that 'zero-rating' internet options do not cater for.[18] ICT developers are rarely from these communities, do not speak their language or understand their problems, and do not easily see money-making opportunities in the community-based management of common resources.

Some of these challenges are not specific to CBNRM systems. Rural users of microfinance products, and some of the farmers who did manage to use ICT solutions to their benefit, are also on the wrong side of digital divides. However, there are two key differences. First, for CBNRM-related ICT solutions to be useful, the users need to consciously contribute much more, in terms of data input, than microfinance users or even digitalised farmers. The latter groups also provide data, but they do this unknowingly, in a black box setting (e.g., their social media activity is analysed and their sensors gather data, but the people involved play no conscious role in this). Second, there are microfinance products and farming apps, such as health insurance and cattle tags, that could cover a large number of people with very limited adaptations, but every CBNRM is uniquely grounded in highly local realities, and ICT solutions would require significant tailoring to fit these realities.

All in all, the detail with which ICT enables humankind to see things going wrong helps generate a sense of urgency, in some quarters, but to date we have not been smart and decisive enough to use the data generated to actually stop deforestation, climate change or air pollution; and until large and overlapping digital divides are overcome, CBNRM systems' use of ICT solutions are bound to be either very basic and generic, or donor-driven, little-used and short-lived.

Notes

1 Shellenberger, M. and Nordhaus, T. (2005) *The death of environmentalism; global warming politics in a post-environmental world*, Strategic Values Project, with the quotation from page 11 (emphasis added). The "large majority" is referring to people in the USA.

2 As defined in Byerlee, D. (9 June 2015) *Growing land scarcity, the Borlaug hypothesis and the rise of megafarms*, International Maize and Wheat Improvement Centre (CIMMYT). (This reference is merely an announcement for a brown bag presentation by Derek Byerlee, but his actual presentation is no longer available online.) The Borlaug Hypothesis is named after Norman Borlaug, who is often presented as 'the father of the Green Revolution' – see, for example, Gillis, J. (13 September 2009) *Norman Borlaug, plant scientist who fought famine, dies at 95*, The New York Times.

3 William Jevons came to this paradox conclusion on the basis of energy efficiency gains in the use of coal, back in 1865. For its relevance in agriculture, see Rudel, T.K. *et al* (8 December 2009) "Agricultural intensification and changes in cultivated areas, 1970–2005", *PNAS (Proceedings of the National Academy of Sciences of the United States of America)*, volume 106, number 49, pages 20675–20680.

4 Hamant, O. (July 2020) "Plant scientists can't ignore Jevons paradox anymore", *Nature Plants*, volume 6, issue 7, pages 720–722.

5 Ofori, B.Y., Owusu, E.H. and Attuquayefio, D.K. (Fall 2012) "Ecological integrity of the Afadjato-Agumatsa Community Nature Reserve (AACNR) of Ghana after a decade of conservation", *Journal of Biodiversity and Ecological Science*, volume 2, issue 4, pages 181–188.

6 The process of disillusionment is described in Dressler, W. *et al* (2010) "From hope to crisis and back again? A critical history of the global CBNRM narrative", *Environmental Conservation*, volume 37, number 1, pages 5–15.

7 See, for example, Kapoor, I. (2002) "The devil's in the theory: a critical assessment of Robert Chambers' work on participatory development", *Third World Quarterly*, volume 23, number 1, pages 101–117.

8 For a comparison between CBNRM and other resource management systems, see chapter 3 of Eekelen, W. van (2020) *Rural development in practice: evolving challenges and opportunities*, Routledge.

9 Blomley, T. *et al* (2016) *REDD+ hits the ground; lessons learned from Tanzania's REDD+ pilot projects*, International Institute for Environment and Development. REDD+ is a voluntary climate change mitigation approach. REDD stands for the 'reduction of emissions from deforestation and forest degradation'; and the '+' refers to the focus on fostering conservation, sustainable management of forests, and enhancement of forest carbon stocks.

10 cbnrm.org/tools/index.html.

11 Chapron, G. (2015) "Wildlife in the cloud: a new approach for engaging stakeholders in wildlife management", *Ambio*, volume 44, pages 550–556.

12 For example, the United Kingdom Butterfly Monitoring Scheme started in 1976 and records data on over 2,000 sites. This figure is from the scheme's website, ukbms.org, accessed on 7 September 2023.

13 Rampheri, M.B. and Dube, T. (October 2020) "Local community involvement in nature conservation under the auspices of Community-Based Natural Resource Management", *African Journal of Ecology*, pages 1–10.

14 See, e.g., Blomley, T. *et al* (2016) *REDD+ hits the ground; lessons learned from Tanzania's REDD+ pilot projects*, International Institute for Environment and Development.

15 Sometimes almost laughably so: a carefully marketed app that was meant to be used by community members to help track monkeys in Panama did not generate *any* data, because *nobody* downloaded the app. Kane, C. and Pitcairn, M. (April 2018) "ICT and community-based primate conservation in the Burica Peninsula", *ENVR 451 Final Research Report*, McGill University, 29 pages, with the fact reported on page 15.

16 In this context it is probably no coincidence that literacy rates and the use of ICT applications are relatively high in the one of the few examples of a successful community based natural resource systems for fisheries: most – 86% – of the Mbenji Island fisherfolk have had some formal education. Haambiya, L., Mussa, H. and Mulumpwa, M. (2020) "A review on the use of information communication technology (ICT) in fisheries management: a case of Mbenji Island small-scale fishery in Malawi", *African Journal of Food, Agriculture, Nutrition and Development*, volume 20, issue 7, pages 17113–17124, with the fact mentioned on page 17117.

17 CICs are cybercafé-type places in rural regions. At one point there were many government-, UN- and NGO-subsidised CICs, generally meant for development purposes but in my experience mostly either empty because they did not actually work, or used for social and entertainment purposes. They were also called 'digital drums', 'LAN houses' (LAN stands for 'local area network') or 'PC Bangs', among other names. Smaller versions, in the form of internet kiosks, have also often been subsidised, and exist under a range of names (e.g., 'wifi kiosks', 'Hello Hubs').

18 'Zero-rating' is the practice of providing limited internet access free of charge, such as Facebook's free but limited FreeBasics. Global Voices reports the languages that FreeBasics offers in six countries (the ones Global Voices selected as its case studies) and it is always English, generally with the addition of a single other language. Global Voices (27 July 2017) *Free*

Basics in real life; six case studies on Facebook's internet "On Ramp" initiative from Africa, Asia and Latin America, Global Voices Advox, page 13. The countries and languages are Colombia (English and Spanish), Ghana (English), Kenya (English and Kiswahili), Mexico (English and Spanish), Pakistan (English and Urdu) and the Philippines (English and Tagalog).

8 The spread of ICT products and services

After the United Nations adopted the *Declaration on the right to development* in 1986,[1] Western donors and NGOs accelerated their shift from needs-based to rights-based and behavioural-change programming. This came at the cost of instant gratification. Donors and NGOs could traditionally feel such gratification when they inaugurated a new water pump or health clinic, or when a new irrigation system was put into use. Rights-based programming rarely provides such photo opportunities. Such programming supports national governments and other 'duty bearers' to meet their responsibilities; and supports a country's citizens and other 'rights holders' to claim their rights. If it works then the results are likely to be long-lasting… but donors and NGOs are less 'in control', the timelines are typically long and many efforts ultimately fail. Behavioural-change programmes also require a great deal of patience, and also often fail to achieve lasting results (see Box 8.1).

And then, when needs-based programmes were no longer *bon ton* but more fashionable programmes were testing the patience of donors and their implementers, modern ICT arrived.

With ICT, some products and services spread at very high speed, and often swiftly and fundamentally changed behaviour. Within the global community of development professionals this led to an uncritical euphoria of the type I described in the prologue and which I experienced myself, for a while. This sense of euphoria was bolstered by our own use of ICT. We are all online, all of the time. Working without our laptop, tablet and smartphone has become almost inconceivable. LinkedIn provides us with an instantly accessible global network of peers, and virtual meetings with participants from around the world are part of the new normal. Most of us use donor-funded resources such as Devex and the Commonwealth of Learning (Devex.com and col.org respectively – both are open-access continuous professional

DOI: 10.4324/9781003451716-10

development resources in the field of socio-economic development). And for our respective fields of expertise we use specialist virtual platforms. Evaluators like me, for example, use and may sometimes contribute to the resources offered by the International Initiative for Impact Evaluation and BetterEvaluation (3ieimpact.org and betterevaluation.org).

Our enthusiasm, fed by success stories and our own utilisation of ICT, translated into a myriad of grants and programmes. Many of them had strikingly specific targets, but actual results were unpredictable and some of the largest successes have come as a surprise. In fact, the most widely used ICT applications in rural regions are social media platforms and free calling applications that were created for recreational uses and without donor funding, and were then *also* adopted for productive purposes. The rapid spread of ICT also led to renewed interest in behavioural economics, and to 'nudge theory' in particular. This theory says that behavioural change is not best achieved with education, legislation or village-square meetings but with timely, subtle nudges provided through channels such as SMS messages, which remind you to "come to the market tomorrow" and "take your pill now".

Box 8.1 ICT investments may have refreshingly swift effects

The benefits of handwashing are enormous and the world is full of programmes that encourage it. However, ample research shows that conventional awareness raising and hygiene education may increase *awareness* but often fail to change actual handwashing *behaviour*.[2] Many years of awareness-raising work about the risks of unsafe sex has led to increased knowledge as well, but again not to very much behavioural change,[3] even in the highest-risk places in the world.[4] Even the seemingly simple problem of people not taking their life-saving medication is hard to overcome[5] (though ICT is helpful here[6]).

In each of these three fields, and in several others, my own evaluations came to similar conclusions, albeit on the basis of far smaller samples. As a consequence of such findings, I – and the international development sector as a whole – had become used to behavioural change interventions that progress slowly, if at all.

Compared to such slow-moving changes, some ICT investments achieved refreshingly swift results. Wherever there is at least 2G coverage, for example, all but the very poorest people are using mobile phones to make calls and send messages.

Part of the spread of ICT is mere show. I recall an Afar man I met in 2010, herding his livestock in full and traditional Afar clothing, with the addition of a mobile phone, hanging around his neck, in a region with no coverage or electricity. He used it as a mere ornament. I have visited many rural schools and offices with computers that were kept in plastic, as they were too precious to unpack and too unfamiliar to actually use. This display of ICT assets as status symbols is not a specifically rural phenomenon. In the early 2000s I was staying in Alexandria, Egypt, when a man was hit by a car and killed, right in front of my apartment, because he had crossed a street without looking. He had been on the phone – except that his 'phone' turned out to be a plastic replica.

Even when people have their phone for use rather than status, many of its opportunities go unnoticed. In 2020, a man in Kenya wrote me to say that he didn't like a book chapter I had written about ICT opportunities in rural regions, because:

> You pretty much only describe internet-enabled services but nobody in rural Africa has internet access. This may sound strange because newspapers and GSMA say that 20–25% of the population has a smartphone. That is correct, but of these people less than 20% buys data – the others use their smartphone only to phone and send text messages. [...] In February I participated in a trip to an area that was only an hour's drive from a provincial city – so not even deep-rural. We spoke with some 100 people, and of these people only one woman had a smartphone [...] and she was unaware that she could use it for anything other than calls and text messages.[7]

This man was right. Online coverage is rapidly expanding throughout rural Africa but this 'technical fix' does not necessarily cause a change in awareness of opportunities, much less behaviour. Even the best-connected African farmers I meet listen to the radio rather than to podcasts, and few of even the wealthier ones would consider the use of drones or remote sensing technologies. In part, this is just because people can't use what they don't know about and in Africa the level of awareness of all but the most basic opportunities is often low. This is why Getaw Tadesse and Godfrey Bahiigwa found that prices fetched by Ethiopian farmers with and without mobile phones were roughly the same: the farmers with a phone just did not know how to use it to access relevant market information.[8] Progress is slow and to this day I sometimes interview organisations that say they opted to disseminate information and engage via radio shows rather than anything more sophisticated, simply because people listen to the radio more than they use other types of ICT. Access to information about ICT possibilities in rural regions is also very unequal and news about opportunities will take longest to reach, say, remotely located illiterate women, or people who rarely leave the farm or cannot move around independently. Chapter 12 on the 'digital divide' explains why, in regions where ICT usage is low in general, the usage among marginalised groups is much lower still.

To increase the use of ICT, governments invest heavily in ICT literacy. In addition, those who sell products seek to increase their appeal and raise awareness of their benefits. In my experience, many such investments are clumsy and generic, and they fail to focus on the specific bottlenecks that hinder a product's uptake. This is unnecessary, as simple surveys could generate insights into the nature of the obstacles. The most commonly used frameworks for such surveys are the *diffusion of innovation theory* and the *technology acceptance model*.

The *diffusion of innovation theory* assumes that a product's uptake largely depends on five things:

1   *Relative advantage*, or the perceived usefulness of a product compared to its alternatives. For people to appreciate a product's usefulness they need to know a certain amount about its features. This does not happen automatically – in fact knowledge and initial interest are often strikingly partial, even for people who have moved beyond phone call and SMS messages. I won't easily forget a man's five-word summary of the benefits of his smartphone: "I use it for porn".[9] This man may be skilled in his particular field of interest, but Esoko's agricultural messaging channels (Esoko.com/content) are unlikely to reach him any time soon.

2   *Compatibility* with the values and needs of individuals and the groups they live in. Traditions, fashion, social pressure, the bandwagon effect and belief systems all matter. They can accelerate ICT's spread but they can also cause a disinterest in things that, to the outsider, seem so very attractive. The most dramatic example I have encountered was an elderly rural woman who rejected an operation that could have reversed her encroaching blindness, and a smartphone that could help her cope (if only because it would bleep when it was time for her pills), because she would rather "surrender to the will of God".

In China, where the government invests heavily in the digitalisation of the economy, uptake is often fast but incompatibility remains a major issue. This is in part because "China is still a traditional culture with strong work and family ethics" and "internet use is [often] perceived as a negative activity that distracts users from valuable life goals".[10] Perhaps counter-intuitively, China's rural population is less hindered by this prejudice than its urban population. Researchers found that:

> As rural residents have much less income and [a] lower living standard than their urban counterparts, they are more influenced by socioeconomic conditions and living environment, and less affected by their perceptions and feelings, in their adoption and use of new media technologies. [In rural areas] behavioural factors, including adoption of other similar media technologies, mass media use, and interpersonal communication, play a more important role than psychological factors in the diffusion process.[11]

3   *Simplicity*. To a large extent, this depends on people's previous exposure to ICT products and services. Provided that this previous exposure had positive results, it is likely to make a new product feel less unfamiliar and intimidating. Unlike many other facets of life, this exposure does not come with age. On the contrary: when it comes to the types of ICT my children grew up with, they find things easy that I find difficult and my parents find impossible.

ANKITA GOT A MICROLOAN!

SOME TRAINING MIGHT ALSO BE USEFUL

New products require new skills

4   *Observability*. If you see others benefiting from an ICT product or service, you are more likely to want to use it yourself. A product might even reach a tipping point that way, when it becomes *so* very common that you are left out if you don't join the club. To illustrate this, the *diffusion of innovation theory* uses a graph that distinguishes between (1) innovators, (2) early adaptors, (3) early majority, (4) late majority and (5) laggards (with the proportion of people on the vertical axis and time on the horizontal one).

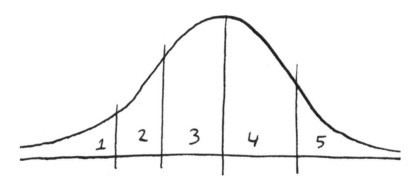

5   *Triability*, which is the ease with which people can try out a new product and explore its features. This is why companies often offer you a free trial month.

The *technology acceptance model* (TAM) is another often-used model that frames questionnaires that seek to identify bottlenecks in ICT uptake. This model focuses on perceptions related to an application's usefulness and user-friendliness.[12] Other models use predictors such as organisational support, perceived risks and social pressure, and the *unified theory of acceptance and use of technology* attempts to combine them all.[13]

For individual products with a well-defined target audience, such surveys often help focus efforts to increase the products' uptake. This is because survey results provide insight into the nature of the obstacles. They may identify a specific type of awareness that needs to be raised or perceptions that need to change (about a product's risks, features or user-friendliness, for example); or a specific skill that needs to be built (basic smartphone literacy or advanced and product-specific feature usage, for example). ICT itself may then be used to remove these obstacles. To reduce forgetfulness, SMS messages may remind potential listeners to "gather your neighbours as the *Voice of the Farmer* radio show starts in 30 minutes", for example, and virtual trouble-shooting facilities may improve a product's user-friendliness and reduce its risks.

Collectively, these survey reports create a map of overall impediments to the uptake of ICT applications. Unsurprisingly, a review of mostly survey-based literature on the adoption of ICT innovations in Africa's agricultural sector "recommends [...] training and empowerment of smallholder farmers to enhance their ability to interact with new agriculture technologies".[14] Countries in other parts of the world have been mindful of this need for some time. In support of the roll-out of key ICT products and systems, countries such as India, China and Indonesia moved far beyond the provision of infrastructure and technical fixes, and invest heavily in perceptions of usefulness, user-friendliness and risks.

Although lack of awareness of the many opportunities and low levels of digital literacy have hampered ICT uptake around the rural Global South, some products are already being used on a massive scale, and sometimes their adoption took very little time. A few products have leapfrogged over previous generations of ICT. Making phone calls, for example, was an entirely new experience in regions that were never connected to landlines, yet mobile phones were quickly used, in Africa and Asia alike,[15] at levels far exceeding the expectations of the providers.[16] They boosted people's social networks and, possibly, entire rural economies,[17] and people quickly identified unanticipated opportunities. These opportunities prominently included hacks to access services or circumvent usage costs. To convey simple, no-cost messages ("I arrived!"), people around the world phone each other and let it ring once before hanging up again. Radio shows use this no-cost option as well, to enable people to vote ("ring once to *this* number if you agree and to *that* number if you do not").[18] When I lived in Egypt, in 1999–2001, a certain type of Nokia phone became particularly popular after somebody discovered that the phone connection stayed, and the call cost nothing, if you removed the SIM card immediately after a connection was made. Once smartphones arrived, illegally downloading software, music and movies became commonplace around the world.

M-Pesa is another example of extreme adoption speed (and, like the BRAC Poverty Graduation Programme covered in Box 5.2, DFID's initial funding for M-Pesa was truly catalytic).[19] Kenya was not the first country to launch mobile payments, but the Kenyan version only required 2G to work, it was easy to use[20] and its uptake was immediate: by December of its launch year (2007) it had reached over 5 million active users, and 75% of all adult Kenyans were using it by 2013.[21] Some agricultural ICT applications also quickly gained momentum: a phone-based climate information service in northern Ghana (which is the poorer part of Ghana), for example, attracted over 300,000 paying subscribers in only two years.[22]

In rural China and in recent years, the uptake of economically useful ICT applications was perhaps faster than almost anywhere else. Moreover, the large gender gap that exists in other parts of the world (only 21% of the 300,000 subscribers of the aforementioned Ghanaian weather app, for example, were women) does not exist in China.[23] The swift uptake in China was possible because of the combination of the rapid spread of internet coverage and the relatively high proportion of men and women owning smartphones.[24] It was also propelled by the rapidly increasing income-earning opportunities available, such as through China's "Internet Plus" rural economy effort,[25] and the strong government encouragement to move to the digital economy. The Chinese government pushed strongly for mobile payments, for example, to formalise the informal economy. When I last visited China, in 2019, it struck me that even the salespeople on the local vegetable markets took their payments through a QR code – at least on the markets in and around Beijing. China's rapid spread of internet connectivity and adoption of smartphones comes with risks and dangers (see Chapter 14) but did increase rural incomes and, relatedly, rural people's subjective well-being, defined as a combination of short-term happiness and longer-term life satisfaction.[26]

One obvious explanation for the rapid uptake of and behavioural change around some ICT products, compared to the slowness of conventional behavioural change, is that they are (1) very clearly and instantly useful and (2) very user-friendly for nearly everybody. These are the two TAM predictors of technology acceptance, and it shows. However, the tangibility and immediacy of the reward and the user-friendliness of the application do not explain every success. Some ICT offerings that serve less tangible and longer-term goals have spread very quickly as well. In a mere few years, the micro-insurance facilities supported by PharmAccess reached nearly 12 million people across Ghana, Nigeria, Kenya and Tanzania[27] – even though, for people unfamiliar with the concept of 'insurance', this is a recurrent expense that does not offer immediate benefits.

Somehow, some ICT products and services instil confidence. People don't know *how* mobile phones and the internet work, exactly, but they see *that* they work and, because the underlying ICT is a black box to the vast majority of users, this observation leads to trust that is extrapolated to other ICT solutions. The first time I saw this blind faith was in Sudan, in 2002, when our cleaning lady, on learning that my wife Maha could use the Internet, asked her to check the true intentions of a potential suitor. More recently, in early 2020, I noticed it when interviewing the owner of a venture capitalist company that uses ODA funding to invest in agritech in Asia. He was a rational and critical man in many ways – but seemed to have blind faith in the accuracy of the algorithms of the start-up agritech companies he was investing in, even though he had not seen them, would not understand them or the subjective choices that underpinned them, and even though these algorithms had not been thoroughly tested yet.

As a consequence of this black-box faith in ICT, many millions of rural people in the Global South embraced some of its innovative products and services, for recreational and productive purposes. People took a number of them to levels, and used them for purposes, that were not anticipated by the companies that invented them. Think of SMS messages, communication apps such as WeChat and WhatsApp, social media platforms such as Facebook and Weibo, or mobile money. Often, people did so without awareness about the risks and dangers that these products and services, and connectivity itself, might pose. Sometimes, these risks 'merely' lead to disappointing results, such as when agritech apps do not in fact help farmers to grow their profits. In other cases, it is worse than that. The second part of this book covers some of the key risks and dangers that ICT poses or contributes to.

Notes

1 UN (4 December 1986) "Declaration on the right to development", *General Assembly Resolution 41/128*, United Nations.
2 This has been researched many times, all over the world. This study is particularly rigorous (it used motion detectors in soap bars, among other things): Biran, A. *et al* (October 2009) "The effect of a soap promotion and hygiene education campaign on handwashing behaviour in rural India: a cluster randomised trial", *Tropical Medicine and International Health*, volume 14, number 10, pages 1303–1314. It says the following, on page 1310: "The intervention achieved a wide reach among the target population and increased reported knowledge of germs. However, at least in the short term, there was no effect on actual handwashing behaviour at key times".
3 John Cleland and Mohammed Ali found that, between 1993 and 2001 – a period with particularly high-intensity safe sex awareness campaigns throughout Africa – results had been very modest: "1.4% annual increase in the reported use of condoms by young single women in 18 African countries". Cleland, J. and Ali, M.M. (2006) "Sexual abstinence, contraception, and condom use by young African women: a secondary analysis of survey data", *Lancet*, volume 368, pages 1788–1793, with the quotations from page 1791.

4 For research on persistent unsafe sex practices in Khutsong, even though people were well-aware of HIV and its causes and of the preventative measures they could take, and even though the HIV infection rate in the 21–25 year age group was 43% at the time of the research, see MacPhail, C. and Campbell, C. (June 2001) "'I think condoms are good but, aai, I hate those things': condom use among adolescents and young people in a Southern African township", *Social Science and Medicine*, volume 52, issue 11, pages 1613–1627.

5 Reasons for non-adherence are many and include stigma and discrimination and side effects, but a review of 154 studies showed that the *most frequently* identified barrier was, quite simply, "forgetting". Croome, N. *et al* (April 2017) "Patient-reported barriers and facilitators to antiretroviral adherence in sub-Saharan Africa", *Aids*, volume 31, issue 7, pages 995–1007.

6 Pop-Eleches, C. *et al* (2011) "Mobile phone technologies improve adherence to antiretroviral treatment in a resource-limited setting: a randomized controlled trial of text message reminders", *Aids*, volume 25, issue 6, pages 825–834.

7 Email in response to the publication of Eekelen, W. van (2020) *Rural development in practice; evolving challenges and opportunities*, Routledge.

8 Mentioned in Chapter 1. Tadesse, G. and Bahiigwa, G. (2014) "Mobile phones and farmers' marketing decisions in Ethiopia", *World Development*, volume 68, pages 296–307.

9 Some people's porn consumption leads to other people's reluctance to use their phones for anything other than calls. For example, Sharifa Sultana and Susan Fussell conducted ethnographic research in three villages in Bangladesh and found that "the villagers stigmatized texting, considering it [a] tool for conducting romantic affairs and sharing porn, so they seldom communicated using text messages." Sultana, S. and Fussell, S.R. (October 2021) "Dissemination, situated fact-checking, and social effects of misinformation among rural Bangladeshi villagers during the Covid-19 pandemic", *PACM on Human Computer Interaction*, volume 5, CSCW2, Article 436, 30 pages, with the quotation from page 7.

10 Nie, P., Sousa-Poza, A. and Nimrod, G. (2017) "Internet use and subjective well-being in China", *Social Indicators Research*, volume 132, number 1, pages 489–516, with the quotation from page 509. Note that this study is limited because it only looked at computer-based internet usage.

11 Wei, L. and Zhang, M. (August 2008) "The adoption and use of mobile phone in rural China: a case study of Hubei, China" [sic], *Telematics and Informatics*, volume 25, issue 3, pages 169–186.

12 TAM was first presented under this term in Davis, F.D. (September 1989), "Perceived usefulness, perceived ease of use, and user acceptance of information technology", *MIS Quarterly*, volume 13, issue 3, pages 319–340.

13 Also known as UTAUT. Explained and tested with strong results in Venkatesh, V. *et al* (2003) "User acceptance of information technology: towards a unified view", *MIS Quarterly*, volume 27, number 3, pages 425–478. This paper also summarises eight distinct ICT acceptance models.

14 Ayim, C. *et al* (2022) "Adoption of ICT innovations in the agriculture sector in Africa: a review of the literature", *Agriculture & Food Security*, volume 11, issue 22, 16 pages, with the quotation from page 12.

15 For an example from Africa, see Aker, J.C. and Mbiti, I.M. (2010) "Mobile phones and economic development in Africa", *Journal of Economic*

Perspectives, volume 24, issue 3, pages 207–232. For an example from Asia, see Wei, L. and Zhang, M. (August 2008) "The adoption and use of mobile phone in rural China: a case study of Hubei, China" [sic], *Telematics and Informatics*, volume 25, issue 3, pages 169–186.

16 Jenny Aker and Isaac Mbiti wrote that "In 1999 [...] the Kenyan-based service provider Safaricom projected that the mobile phone market in Kenya would reach three million subscribers by 2020. Safaricom, alone, currently [i.e., in *2009*] has over 14 million subscribers". see Aker, J.C. and Mbiti, I. M. (2010) "Mobile phones and economic development in Africa", *Journal of Economic Perspectives*, volume 24, issue 3, pages 207–232, with the quotation from pages 209–210.

17 The Economist claims that "every 10% increase in mobile-phone penetration in poor countries speeds up GDP growth per person by 0.8–1.2 percentage points a year", though it is unclear what research this claim is based on. The Economist (11 November 2017) *Beefing up mobile-phone and internet penetration in Africa; without connectivity, nothing moves*, The Economist. (This is the online title; in the print version of the Economist the article is titled "The right connections".)

18 Farm Radio International (FRI), for example, uses this and calls it 'beep-2-vote' technology. Hudson, H.E. *et al* (2017) "Using radio and interactive ICTs to improve food security among smallholder farmers in Sub-Saharan Africa", *Telecommunication Policy*, volume 41, pages 670–684, with the fact reported in Table 1 on page 673.

19 The original DFID programme document no longer seems to be publicly available, but the development of M-Pesa was financed with a grant from DFID's Financial Deepening Challenge Fund, as mentioned in the original press release of Safaricom and Vodafone, which is available in Annex 2 on page 42 of Arunachalam, R.S. (May 2007) "Microfinance and innovative financing for gender equality: approaches, challenges and strategies", *Eighth Commonwealth Women's Affairs Ministers Meeting, Kampala, Uganda, 11–14 June 2007*, Commonwealth Secretariat.

20 For a concise overview of M-Pesa's usability, see Khan, F. (30 April 2017) *Why is M-Pesa the posterchild for financial inclusion?*, LinkedIn (The issue of usability is not only covered in the brief section titled "usability".)

21 Onsongo, E. and Schot, J. (2017) "Inclusive innovation and rapid socio-technical transitions: the case of mobile money in Kenya", *SPRU Working Paper Series*, 2017–07, pages 1–28, with both facts mentioned on page 2.

22 Partey, S.T. *et al* (April 2019) Scaling up climate information services through public–private partnership business models; an example from northern Ghana, *CGIAR Info Note*, Consultative Group for International Agricultural Research.

23 For specifics on the virtually non-existent gender digital divide in rural China, see CIW (April 2020) *Statistics: China internet users*, China Internet Watch.

24 *Ibid*. Internet penetration in rural China increased from 28% in 2013 to 46% in March 2020 (compared to 60% and 77% in urban areas), and by March 2020 over 28% of the rural population *used* internet – almost invariably through smartphones (against 72% in urban areas; no 2013 figures available).

25 Described and assessed in ADB (September 2018) *Internet Plus agriculture; a new engine for rural economic growth in the People's Republic of China,* Asian Development Bank.
26 Nie, P., Ma, W. and Sousa-Poza, A. (January 2020) *The relationship between smartphone use and subjective well-being in rural China,* Electronic Commerce Research.
27 According to page 22 of PharmAccess Group (2020) *Progress report 2019,* PharmAccess Group (the stated figure is 11.8 million people).

Part II

The risks, dangers and externalities of ICT

Part I covered a range of ways in which ICT applications seek to mitigate against risks and dangers. Once fully functioning, TartanSense's mini robots recognise and remove diseased cotton plants on small farms, and this reduces the risk of epidemics and significant harvest losses. Pix Fruit, the app that uses big data to help a farmer estimate the volume of the harvest when it is still on the trees, reduces the risk of trade deals that are based on wildly inaccurate guesstimates. Micro-insurance mitigates against the risks of harvest failure and poor health. Social assistance programmes that use SIM cards face lower risk of theft than programmes that distribute actual cash, as there are no cash depots to raid, wallets to rob or 'fees' to pay to the person distributing the money. Online driving test applications and virtual driving tests reduce the chance of bad drivers on the road. And so forth.

Whilst ICT may reduce risks and mitigate their effects, ICT also sometimes *poses* or *contributes* to risks and dangers. When these risks and dangers materialise, both recreational and productive uses of ICT applications may destroy lives, livelihoods and liberties. These risks and dangers are often unanticipated, and they are *hard* to anticipate because technological innovations often have consequences that are dramatic but largely unforeseen.

This second part covers ICT-related risks and dangers related to:

- Farming and rural employment
- Financial safety and security
- Gambling
- The effects of unequal access to ICT products and services
- Misinformation and disinformation
- Inclusivity and civic space.

DOI: 10.4324/9781003451716-11

Each of these risks and dangers is worthy of attention because of its far-reaching implications for the people, groups and societies involved. However, other choices would have been possible. The issue of gambling, in particular, is one of several possible case studies of harmful 'recreational' use of ICT. It would have been possible to focus, instead, on the emerging finding that long hours of smartphone use reduce people's subjective well-being,[1] or on the effects of cyber bullying.[2] I chose to focus on ICT-powered gambling because it presents a juxtaposition with remittances and social assistance. All three are financial flows occurring outside of the trade economy, but while the remittance and social assistance flows move largely from urban centres to the rural Global South, ICT-powered rural gambling generates financial flows that move in the opposite direction.

Notes

1 Nie, P., Ma, W. and Sousa-Poza, A. (January 2020) *The relationship between smartphone use and subjective well-being in rural China*, Electronic Commerce Research, with the fact covered in Table 5 and on page 17 (of 28 – the pages are not numbered); Keles, B., McCrae, N. and Grealish, A. (2020) "A systematic review: the influence of social media on depression, anxiety and psychological distress in adolescents", *International Journal of Adolescence and Youth*, volume 25, number 1, pages 79–93.

2 Google Scholar gives 13,700 hits with 'cyberbullying' in the title (on 7 September 2023). Almost no research is situated in the rural Global South, but the studies that do exist come to a common conclusion: it is a problem, and it often takes the shape of revenge porn. See, e.g., Farhangpour, P., Mutshaeni, H.N. and Maluleke, C. (2019) "Emotional and academic effects of cyberbullying on students in a rural high school in the Limpopo province, South Africa", *South African Journal of Information Management*, volume 21, issue 1, 8 pages.

9   Farming and rural employment

ICT applications help farmers in various stages of the agricultural production process. However, they do this mostly for a few crops that are of key importance for the global agro-industry, and generally not for the vast majority of crops that the industry finds to be of lesser economic weight. TartanSense focuses on cotton. Pix Fruit focuses on mangos, with plans to expand into coffee, lychees and citrus fruits. Babban Gona focuses on corn and rice. There are no comparable apps for spider plant, mongongo, moringa, cherimoya or fonio. This reinforces the agribusiness's blind spot for indigenous crops and leads to a 'precision divide'[1] between a modest number of widely-planted crops and everything else. In the short run, this may reduce the diversity of a farming community's income sources, as most farmers are cultivating the same small number of key crops. In the long run, this may reduce the world's overall agrobiodiversity.

Another problem is that, for crop-specific and crop-agnostic ICT applications alike, advice might not be *good*, and bad advice is bad advice even if the slick presentation of modern-day ICT applications makes it look convincing. Bad advice may be given where ICT merely facilitates communication. Vets communicating via WeChat are the same vets that used to make farm visits, but they now base their advice on mere pictures and phone calls instead of face-to-face visits and physical assessments of livestock. Another risk is that their virtual advice may not be trusted in the same way that face-to-face advice is trusted, and "research across different disciplines emphasizes that trust in the information source affects the way information is interpreted, accepted, and acted upon".[2]

Bad advice may also be given on the basis of data-driven algorithms. At the time of my visit to TartanSense, the self-learning algorithms upon which the pilot robots based their actions were not yet good enough, and the pilot cotton farms using them were knowingly taking a risk. More generally, agricultural big data analytics is relatively new and still under development, and the trade-off between accuracy and speed (important in the context of weather and market prices, for example) may further compromise the quality of the output.[3] Moreover, even non-time-sensitive products of well-established agritech companies may give bad advice. This

DOI: 10.4324/9781003451716-12

advice may ultimately be based on inaccurate or incomplete data (not to be confused with an insufficient *volume* of data), on data that cover too short a period (also because many agritech products are still relatively new) and on flawed algorithms. In traditional agronomic research, an analysis can be, and sometimes *is*, checked for typical shortcomings, such as "misinterpreting supposed signs [...], irresponsible guesswork and upscaling, [...] interpolating trends between points that followed from random variation, and [...] ignoring the spatial and temporal variability in farming methods".[4] It is nearly impossible to conduct such checks if an analysis is based on hidden data and opaque algorithms. In such cases, the only way to verify the quality of the analysis and concomitant advice is for farmers to *field*-test them.[5] This lack of checks is risky, and often based on the default assumption that these digital data measurement and analytical tools must be correct. Adverse outcomes are often seen to be the result of "incomplete adoption and incorrect use by farmers, or [...] temporary technical 'bugs' that will rapidly be 'fixed'". This "exaggerated belief in the precision of big data [...] over time leads to an erosion of checks and balances (analogue data, farmer observation et cetera) on farms".[6]

Even if advice is good in and of itself, the collective system of app-based advice and 'nudges' may eventually lead to an erosion of indigenous knowledge and a deskilling of farmers if the advice of apps takes the place of what used to be "a dynamic, hybrid, group process integrating environmental and social learning, in which farmers observe, discuss, and often participate in each other's operations".[7] In well-connected regions where people have smartphones, it often strikes me that the sheer amount of time young people spend at home and online, rather than in the fields or forests and socialising with local peers and elders, must erode the experiential learning and social dynamics through which indigenous knowledge is shaped and exchanged. Jason Young saw this among Canadian Inuit, and quotes a man who gives a typical Inuit example of a universal phenomenon:

> One time we were gonna go caribou hunting and we heard that one of our friends [...] was going to go caribou hunting with us. But, when we left he was a no-show. [We] asked him why he didn't come ... he said he had everything packed, he just had to go and put on his warm clothes and go. But he started watching a movie and stayed home.[8]

Fiddling with phones and earbuds during work may also numb one's observation skills somewhat and this, too, is something Young noticed among the Inuit he studied:

It quickly became evident that Inuit are training their attention on technology even when out on the land in the Arctic. In many instances this is leading Inuit to pay less attention to their surroundings.[9]

In addition to posing risks to individual farmers, hunters and others who rely on farming or nature's off-farm riches, ICT may widen existing inequalities, for three reasons:

- ICT may adversely affect the opportunities and relative position of low-skilled rural labourers.
- Larger farmers may gain yet more advantages over smaller farmers.
- ICT often equalises access to information and, with that, tilts the power balance between agro-companies and farmers towards the latter – but parts of ICT, and big data analytics in particular, often have the opposite effect.

The next few pages cover these three issues in turn.

Issue 1: low-skilled rural labourers are typically among the poorest groups in rural society, and they may lose out where ICT substitutes labour. Pesticide-spraying drones and remote irrigation could cause landless labourers to lose part of their income, to give just two of many examples of where this may happen. My own evaluations were inconclusive about the net effects of such labour-saving ICT applications, as I saw both labour lost and labour gained, but less anecdotal research concludes that the net effect is often labour lost. On the basis of a variety of surveys and databases, even the otherwise strongly pro-ICT 2016 "World development report" concluded that "from a technological standpoint, two-third of all jobs are susceptible to automation in the developing world", although "the effects are moderated by lower wages and slower technology adoption".[10] Several other multilateral organisations (e.g., ILO, DESA, OECD, WEF) and institutions (Citi GPS, McKinsey Global Institute) come to comparably large estimates.[11]

Such labour substitution is not necessarily a problem, as the labour market of a healthy economy tends to adapt and seize new opportunities. In the mostly Asian countries in which the Green Revolution was successful, this revolution also led to a certain amount of labour substitution, but the net effects were nonetheless positive for rich and poor alike, because the productivity gains set in motion a positive spiral of rural–urban migration, economic development, industrialisation and rural economic diversification (see Box 9.1). A new ICT-driven wave of labour substitution could again accelerate economic diversification in

Asia, which could again absorb labourers whose original jobs are taken over by robots and other labour-saving tools – especially where the poorest rural segments gain literacy and ICT skills.

Box 9.1 Substituting labour may create a positive spiral of economic development

Nowadays, most researchers on the 'Asian miracle' agree that the process of rapid economic growth in much of Asia, in the last few decades of the 20th century and beyond, was sparked by heavy initial investments that enhanced agricultural production.[12] These investments started a positive spiral, in non-linear processes that differed across countries but broadly looked as follows:

Small-scale farmers adopted better seeds, and cultivated them with heavier use of machinery, irrigation, fertilisers, pesticides and herbicides. Governments facilitated this transition, by subsidising and promoting these new farming methods; by guaranteeing minimum crop prices; and through investments in irrigation and better rural roads. Governments also facilitated the expansion of agricultural lands and pursued land reform that redistributed land from wealthier to poorer farmers, as well as land consolidation that re-divided plots into more efficient plots. Together, these various types of support and new practice increased agricultural production, and the productivity per farm worker. This led to higher earnings among farmers and their workers, whose health and nutritional status improved as a consequence. This enabled them to utilise farming opportunities more fully, which had positive effects on production, and so forth. The impact of this Green Revolution, on individual farmers and on large parts of Asia (and on some countries in other regions, especially in Latin America) was dramatic. It turned some Asian countries from net importers to net exporters of rice; helped avoid a massive famine in India and reduced malnutrition elsewhere; and explains most of the rapid expansion of agricultural production in many Asian (as well as a few Latin American) countries, between the late 1960s and today.[13] This increased agricultural production reduced the price of some local foodstuffs, which benefited non-farming households. In some regions it increased the price fetched for export crops, as the increased crop volume made these regions more attractive to international buyers.

The higher agricultural incomes created a higher demand for education and health services, which then created a positive spiral as income, health and education all reinforced each other. The higher incomes also caused rural economies to diversify, as demand for non-farm products increased. This diversification meant that people's economic activities were no longer confined to farming. Though farming was much more profitable than it had been before, millions of people moved to an ever-expanding range of occupations that catered for a booming demand for, say, fixing motorbikes, organising weddings and selling mobile phones. Millions of people moved to the city, which facilitated industrialisation, which accelerated economic growth, which reinforced both the demand for rural products and investments in rural development. As this rapid development occurred in several countries in the region, in the same few decades, it created regional momentum (i.e., a positive 'neighbourhood effect'), whereby simpler types of exportable economic production gradually moved from more economically and technologically advanced countries, with higher salaries, to countries that were lagging behind and paid lower salaries.

However, the rural–urban migration *in Africa* often follows a pattern that is close to the one assumed in the 'Lewis model'.[14] In this model, the marginal labour productivity in agriculture is zero and rural economic diversification is limited. In such conditions, rural–urban migration is not the consequence of meaningful income-earning opportunities in the city but of an absence of income-earning opportunities on the farm or in the village. The introduction of labour-saving ICT in such conditions reduces the number of labourers beyond which additional workers have negligible productivity. In such circumstances, ICT may generate economic growth, but it is inequitable growth that leaves the poorest segments of society worse off. This is compounded by the fact that Africa cannot hope to follow Asia's 20th-century model of labour absorption through the rapid growth of initially labour-intensive manufacturing industries – partly because international manufacturers often do not find African investment destinations sufficiently attractive,[15] and partly because changes in production technology and consumer tastes mean that the global manufacturing industry no longer has sufficient labour absorption capacity.

Non-agricultural work in rural economies is also at risk, because "the ability of farming to stimulate the RNFE [rural non-farm economy] depends greatly on [...] the social and spatial organisation of agricultural value chains [...]".[16] An app such as Khula, which creates 'one big virtual farm' that procures inputs at low and sells harvests at high prices, fundamentally changes the spatial organisation of agricultural value chains, as such apps help farmers, including smaller-scale farmers, to access more distant markets to optimise their profits. Such apps may improve a farmer's income, but also reduce the liveliness of income-generating markets close-by. Similarly, apps like OneBridge, that enable rural people to make online purchases, vastly increase the product range available to rural people but also reduce local retail jobs. These jobs are replaced by jobs in the OneBridge distribution hubs, which are urban jobs, and by the jobs created through its last-mile delivery infra-structure, which are also generally urban, as the people delivering the packages start and end their day at the distribution hubs.[17]

The loss of farm and off-farm work opportunities has a ripple effect in local economies. In *Rural development in practice*, I summarised an assessment I had conducted of the effects of a new weekly local livestock market in the rural Afar region of Ethiopia:

> This new local livestock market had made life [for local pastoral-ists] easier and profits higher. The sales volume increased every week, and buyers and sellers were willing to travel increasingly long distances. Indeed, some meat factories from other regions now sent their representatives, who arrived on the Friday and left on the Sunday. When I visited this market, only a year after it had opened, it already featured more than livestock. You could now buy agri-cultural produce and animal derivatives, like sheep wool and camel fat (which people use as hair gel), as well as textiles and a range of other manufactured products. The surrounding service industry had also benefited: a restaurant owner now had more customers and spent much less time searching for the sheep and goats she needed for her dishes, because she could buy them at the market.
>
> New businesses had opened. Somebody bought a few mattresses, called his house a 'hotel' and offered the traders a good night's rest. Somebody else took to tanning and selling hides. An entrepreneurial football fan bought a solar-powered battery, a satellite receiver and a television, and was making money by inviting traders and others to watch the English Premier League in his yard.[18]

If e-commerce and 'one big virtual farm' concepts maintain their current momentum, they will significantly reduce the ability of such local markets to generate these types of local economic multiplier effects.

Issue 2: poorer farmers lose out if ICT solutions help wealthier farmers to increase their market share. Compared to poorer farmers, wealthier farmers are better able to make innovative investments. They have more money to spend and are more likely to be able to borrow money, even if microfinance developments described in Chapter 4 are reducing the gap. The implication is that only the wealthier farmers adopt the costlier ICT solutions, such as the use of digital soil maps, drone imagery and computer-operated greenhouses. If indeed the ICT applications do as they promise, this causes their advantage over smaller farmers to increase as the latter group is left behind with, in relative terms, poorer crops that require higher levels of inputs.

Other ICT opportunities do not need much money. Smartphones and internet are increasingly affordable and many of today's farm-focused apps are free or almost free of charge. Even in such cases wealthier farmers may benefit disproportionately: research conducted among farmers in China concludes that "individual income positively determines farmers' decision to adopt the ICTs".[19] First, they may jump at opportunities before the poorer farmers even know they exist, as they are likely to have wider networks that alert them to possibilities, and they are likely to be more literate and tech-savvy. Second, even if both types of farmers have access to, say, the same apps that provide real-time information about market opportunities, these opportunities might help the wealthier farmers and not the poorer ones. This is the case if, for example, the wealthier farmers have quicker access to useable roads that reach that market, or access to motorised transport, or if they are the only farmers who are able to sell at the minimum volume required. In such cases, and compared to the traditional method of going to a physical market and hoping for the best, the wealthier, better-connected farmers win and the poorer ones lose.

Issue 3: there are ICT applications, and big data applications in particular, that benefit agro-companies at the cost of farmers. Chapter 2 discussed how market-related ICT improves farmers' access to information, and how this potentially reduces the power imbalance vis-à-vis other market players. This is often true, but it is only part of the story and there are also situations where ICT benefits transnational agro-companies (mostly rooted in the USA, Europe or China) at the cost of farmers – especially where farmers cultivate crops that are part of long global value chains.

This could happen when companies use big data for unethical exclusion, such as by using predictive satellite imaging to identify farmers whose harvests are likely to fail and excluding them from insurance. It could also happen through targeting (e.g., by using call logs and social network data to identify vulnerabilities that are then used to push loan offers) or negotiations (e.g., by using ICT-powered regional harvest forecasts to negotiate prices before farmers know that crop market prices are likely to be high). It may also happen where companies create dependencies. Unless companies transfer farm-specific big data to new product providers when farmers decide to switch providers, farmers cannot leave companies without losing the data required to optimise their farming. Mark Ryan presents a case study of a company that does provide such data transfer services,[20] but I have not come across any such examples myself, and it is rarely enshrined in enforceable legislation.[21] Companies also create dependencies when they 'black-box' their products, so that they alone are able to conduct maintenance operations or fix problems;[22] or when their products require an upfront investment that is company-specific. One company I visited provided spraying robots to small-scale farmers, which required product-specific sachets. The company owners referred to its 'lock-in' practice as their 'Nespresso model', and it was key to their business model. These dependencies are a little like the dependencies in tenant farming, where farmers are in an unequal and inescapable income-splitting relationship with landowners – but in the case of big data it is not the landowner but the data owner that takes the cut, and it takes a few years before the farmers find themselves stuck in the relationship.

Big data consolidates inequalities

Notes

1 A term coined by Visser, O., Sippel, S.R. and Thiemann, L. (August 2021) "Imprecision farming? Examining the (in)accuracy and risks of digital agriculture", *Journal of Rural Studies*, volume 86, pages 623–632.

2 Aker, J.C., Ghosh, I. and Burrell, J. (2016) "The promise (and pitfalls) of ICT for agriculture initiatives", *Agricultural Economics*, volume 47, supplement, pages 35–48, with the quotation from page 37.

3 Zhang, H. *et al* (2015) "Agriculture big data: research status, challenges and countermeasures", in Li, D. and Chen, Y., editors, *Computer and computing technologies in agriculture VIII*, Springer, pages 137–143. ('Computer and Computing Technologies in Agriculture', or CCTA, is a conference, and these were the proceedings of the eighth one – hence the VIII in the title.)

4 Niek Koning and Eric Smaling presented this list of potential shortcomings in the context of soil erosion research in particular. Koning, N. and Smaling, E. (2005) "Environmental crisis or 'lie of the land'? The debate on soil degradation in Africa", *Land Use Policy*, volume 22, number 1, pages 3–11, with the quotation from page 4.

5 Visser, O., Sippel, S.R. and Thiemann, L. (August 2021) "Imprecision farming? Examining the (in)accuracy and risks of digital agriculture", *Journal of Rural Studies*, volume 86, pages 623–632.

6 *Ibid*, with the quotations from page 623 and the abstract respectively. Pro-ICT lobbyists such as the World Bank reinforce this bias with optimistic references to the good things that "objective measurement methods" bring farmers (for one of many examples and a reference to these 'objective measurement methods', see World Bank (2021) *World development report: data for better lives*, A World Bank Group Flagship Report, pages 29–30). This precision bias reinforces a more general 'quantification bias' that causes big data to trump qualitative 'thick data' in people's minds. See, e.g., Wang, T. (28 November 2016) *The cost of missing something*, TEDxCambridge, YouTube, starting on the tenth minute.

7 Brooks, S. (2021) "Configuring the digital farmer: a nudge world in the making?", *Economy and Society*, volume 50, issue 3, pages 374–396, with the quotation on page 387.

8 Young, J.C. (2019) "The new knowledge politics of digital colonialism", *EPA: Economy and Space*, volume 51, number 7, pages 1424–1441, with the quotation on page 1434.

9 *Ibid*, with the quotation on pages 1435–1436.

10 World Bank (2016) *World development report 2016; digital dividends*, A World Bank Group Flagship Report, with the quotation from Figure 2.24 on page 129, and the full references to the sources underneath that figure.

11 These various estimates are summarised in Table 1 on page 19 of ILO (2019) *Work for a brighter future*, Global Commission on the Future of Work, International Labour Organization.

12 E.g., Henley, D. (2015) *Asia–Africa development divergence: a question of intent*, Zed Books; Koning, N. (2017) *Food security, agricultural policies and economic growth; long-term dynamics in the past, present and future*, Earthscan; Gollin, D., Hansen, C.W. and Wingender, A. (2018) "Two blades of grass: the impact of the Green Revolution", *NBER Working Paper 24744*, National Bureau of Economic Research; and Timmer, C.P. (2015)

Food security and scarcity: why ending hunger is so hard, University of Pennsylvania Press and the Center for Global Development.

13 The Green Revolution is generally seen as a phenomenon of the 1960s and 1970s, but agricultural productivity gains in large parts of the Global South actually accelerated *after* that. See Fuglie, K. (2015) "Accounting for growth in global agriculture", *Bio-based and Applied Economics*, volume 4, issue 3, pages 201–234 (with the point I'm making here aptly summarized in conclusion 3 on page 224).

14 Lewis, W.A. (May 1954) "Economic development with unlimited supplies of labour", *The Manchester School*, volume 22, issue 2, pages 139–191.

15 This opinion may sometimes be grounded in stereotyping and unfamiliarity but also simply reflects realities that are quantified and specified by country in the 'ease of doing business' index (see data.worldbank.org/indicator/ic.bus.ease.xq?view=map, with lighter colours reflecting easier business climates).

16 Du Toit, A. (2016) "Can agriculture contribute to inclusive rural economies? Findings from southern Africa", *PLAAS Policy Brief*, number 45, with the quotation from page 1. (The other two key dependencies Andries du Toit mentions, in this context, are the scale of agriculture and the political economy of local institutions.)

17 When I interviewed a group of OneBridge delivery men they all turned out to have rural roots but currently urban residence.

18 Eekelen, W. van (2020) *Rural development in practice: evolving challenges and opportunities*, Routledge, with the quotation from page 108.

19 Zhu, Z., Ma, W. and Leng, C. (2020) "ICT adoption, individual income and psychological health of rural farmers in China", *Applied Research on Quality of Life*, 21 pages, with the quotation from page 16 and details on page 9.

20 Ryan, M. (2019) "Ethics of using AI and big data in agriculture: the case of a large agriculture multinational", *The ORBIT Journal*, volume 2, issue 2, pages 1–27, with a description of the problem on pages 9–10 and the example of the company confirming that "the farmer owns their own data and they can move to a different farm management system supplier, with that data, if they choose to" on page 18.

21 A problem outlined and then applied to South Africa on pages 24–25 of Aguera, P. *et al* (June 2020) *Paving the way towards digitalising agriculture in South Africa*, Research ICT Africa.

22 Another method companies use to ensure they are in control of maintenance and repair of their equipment is to expressly forbid owners to do this themselves or to contract anybody other than authorised repair agents to do this. I do not know how common this already is, in the Global South. See, for example, Hirsch, J. (August 2019) *As farmers fight for the right to repair their tractors, an antitrust movement gains steam*, The Counter.

10 Financial safety and security

ICT may prevent, reduce or expose scams, fraud and corruption, but it also presents new opportunities for just that. At the lowest level, this takes the shape of tricking people into sharing the details of their mobile wallets.[1] At higher levels, corruption risks are partly just another manifestation of 'normal' corruption vulnerabilities that large procurement contracts are subject to in highly corrupt countries. In this context, and with Nigeria serving as a predictable case study, Peerayuth Charoensukmongkol and Murad Moqbel found a U-shaped correlation between a government's ICT investments and corruption. They posited that ICT investments often reduced corruption but that, in some countries, *overinvestments* in ICT were themselves corrupt investments, of the type of a million-dollar contract for a thousand-dollar satellite dish.[2]

More specific to the nature of ICT itself, e-government may help reduce corruption (see Chapter 6), provided there is a minimum level of institutional quality and ICT literacy within the country.[3] In the absence of that, ICT may instead *facilitate* corruption. I have come across a few such ICT-facilitated cases of corruption. I cover one in Box 10.1, below, from when I worked in Bosnia and Herzegovina. I saw another example in Uganda, where programme staff and payment agents defrauded recipients by embezzling SIM-based back payments.[4] The latter example illustrates a phenomenon that Seifallah Sassi and Mohamed Sami Ben Ali describe for Africa as a whole: "ICT diffusion may create new opportunities for bribery in countries where the threshold of rule of law has not yet been reached".[5]

Fraud is not only committed by corrupt government officials and their agents. Many *yahoo boys* also scam people, using online smartphones and mobile phones. These *yahoo boys* are typically from poor (urban) backgrounds and they are often glamorised, in celebratory songs and poems, for scamming money out of the domestic elite and people from

DOI: 10.4324/9781003451716-13

the West.[6] These songs and poems do not mention that they target poor people as well, but they do. As David Medine summarises:

> In the Philippines, Peru, India, Kenya, South Africa and many other developing countries, poor people [...] have been targeted by online fraudsters. [...] There is the risk that these scams could undermine confidence in digital technologies that are proving so very important in keeping people informed and connected during the pandemic. In particular, trust in digital financial services, which have been useful in advancing financial inclusion efforts, could be damaged at the very time that they have proven to be an effective means of getting payments to poor people quickly and efficiently.[7]

I have looked for but not found any research on the extent to which online and mobile phone scams penetrate into rural communities, and I have not come across such scams myself – but I cannot think of a reason why rural communities would be shielded from this risk.

Box 10.1 E-government sometimes facilitates corruption

During my employment at the Office of the High Representative in Bosnia and Herzegovina (OHR), we exerted pressure on the Bosnian authorities "to take whatever steps are necessary to [...] resolve all outstanding legislative issues with regard to the CIPS [the 'Citizen Identification Protection System'] project [and] to increase the level of management and budgetary resources needed to implement the CIPS project, given the importance of this project to the domestic and international security interests of the citizens of BiH [Bosnia and Herzegovina]". CIPS was piloted in the first half of 2002 and was to be "a major step forward in safeguarding [of] individual identity".

CIPS did not achieve what it set out to do, and "caused a great deal of damage to the reputation of the public service in the whole country", because:

> Police officers used their official computers and authorities to enter the system and change data. Mostly, they would find a person in the database who had BiH citizenship yet had never received a personal ID card (for example, if the person went abroad during the war and never came back). They then sent the person who wished to receive a false document to a registrar's office (another part of the organised criminal group operating within the

jurisdiction of the municipal administration) where they would provide a birth certificate and a certificate of citizenship in a false
name, which was enough to start the procedure for obtaining the
personal ID card. Even the data of dead persons was used; rather
than officially registering the death, they registered the person as
having lost an existing, valid ID card and started the procedure for
the issuance of a new ID card.

Sources: The first two quotations are from, respectively, OHR (28
June 2002) *High Representative amends CIPS laws to meet CoE and
international standards*, Office of the High Representative; and PIC
(20 March 2001) *Communique by the PIC Steering Board*, Peace
Implementation Council. The last two quotations are from pages 42
and 41, respectively, of ReSPA (2013) *Abuse of information
technology (IT) for corruption*, Regional School of
Public Administration.

Notes

1 I have seen people sharing their details a few times, and Priscilla Twumasi
Baffour and her colleagues conclude that "the unsafe practice of sharing m-
wallet details with third parties remains the main source of risk". Baffour, P.
T., Abdul Rahaman, W. and Mohammed, I. (2020) "Impact of mobile money
access on internal remittances, consumption expenditure and household welfare in Ghana", *Journal of Economic and Administrative Sciences*, volume
37, number 3, pages 337–354.
2 Charoensukmongkol, P. and Moqbel, M. (March 2014) "Does investment in
ICT curb or create more corruption? A cross-country analysis", *Public
Organization Review*, volume 14, number 1, pages 51–63.
3 For ease of reference again: Adam, I.O. (May 2020) "Examining e-government development effects on corruption in Africa: the mediating effects of
ICT development and institutional quality", *Technology in Society*, volume
61, 10 pages.
4 Back payments are payments that were initially delayed and then combined
with the next payment. Such delays are common in cash transfer programmes in several countries in the Global South. See ICAI (January 2017)
*The effects of DFID cash transfer programmes on poverty and vulnerability;
an impact review*, Independent Commission for Aid Impact, Table 3 on page
25. The example of Uganda is covered in DFID Uganda (revised version,
October 2015) *Expanding social protection in Uganda – phase II, 2015–16–
2019–20; business case*, Department for International Development, footnote
5 on page 9.
5 Sassi, S. and Ali, M.S.B. (2017) "Corruption in Africa: what role does ICT
diffusion play" *Telecommunications Policy*, volume 41, pages 662–669, with
the quotation on page 667.

6 See, e.g., Warner, J. (January–July 2011) "Understanding cyber-crime in Ghana: a view from below", *International Journal of Cyber Criminology*, volume 5, number 1, pages 736–749; Tade, O. (July–December 2019) "Cybercrime glamorization in Nigerian songs", *International Journal of Cyber Criminology*, volume 13, issue 2, pages 478–492. Nigeria is a country with a particularly high number of scammers, and it tops the list in John, A. (undated) *10 countries with the highest number of scammers*, Wonderslist.

7 Medine, D. (April 2020) "Financial scams rise as coronavirus hits developing countries", *CGAP blog series on coronavirus (Covid-19): financial services in the global response*, Consultative Group to Assist the Poor.

11 Gambling

One of the negative side effects of financial ICT innovations is that they may cause financial despair. M-Pesa is a case in point. M-Pesa became the poster child for rapid financial inclusion and was much celebrated and promoted, especially after the publication of a paper by Tavneet Suri and William Jack, published in *Science*, which attributed large-scale poverty reduction to M-Pesa's mobile money.[1] This claim was often used by the World Bank and other stakeholders working in the field of financial inclusion.[2] The problems mobile money proponents did mention were put down to mere technical glitches. Failed transactions or delayed confirmation messages were attributed to congestion during peak texting times, and an inability to pay out on the side of rural M-Pesa agents was the consequence of cash flow constraints, for example.[3] However, Suri and Jack's underpinning assumptions were refuted in 2019.[4] Since then, more negative research has appeared, often about unregulated mobile money providers causing indebtedness that reinforces poverty. Over 2.7 million Kenyans are blacklisted after digital loan defaults,[5] for example, and the April 2019 "FinAccess Household Survey" of Kenya[6] presents a bleak picture of the effects of digital microfinance. As summarised by Sibel Kusimba:

> This extensive survey of consumer finance and digital finance showed that in spite of the greater reach of finance and credit products into people's lives, their overall financial health had greatly declined since 2016. Only 20% of Kenyans – all income levels – were financially healthy, as measured by their ability to save for emergencies, plan, and meet their daily needs (down from 39% in 2016). Many had lost money to digital scams. Although inclusion in terms of accounts and usage of products has increased, people are not benefiting. It is hard to reconcile the celebratory narrative of M-Pesa with the consumer risks it soon created.[7]

DOI: 10.4324/9781003451716-14

This survey differentiated between rural and urban areas, and found the situation in rural areas to be far worse than in urban ones: only 14% of rural people were in good financial health, compared to 33% of urban people.[8]

A large but little-known part of the indebtedness is not caused by investments going wrong but by online and other forms of ICT-powered gambling. Gambling is within the realm of the 'recreational use' of ICT applications. Recreational use is the world's most common type of individual ICT use, but it and its risks and dangers are widely ignored by researchers within the field of rural development and by the aid industry alike.[9] This blind spot applies to online gambling as well. I have not heard of research on the parallels between the recent surge in online gambling and the equally sudden rise in pyramid and Ponzi schemes in former communist countries, in the early years after the end of the Cold War, for example.[10] Similarly, I have searched for but have not seen a single pre-2019 donor paper or academic publication that predicted that mobile money might create a problem of indebtedness by suddenly exposing an unprepared population to the heavily advertised prospect of a new way of making money through gambling. Online gambling was a 'Black Swan' – a phenomenon that was easy to explain after its existence had become obvious, but an unanticipated and suddenly life-changing phenomenon in its early stages. As with the pyramid and Ponzi schemes in Eastern Europe, new forms of gambling, powered by the instant link between betting options, mobile money wallets and microloans, is likely to cause major setbacks in the financial position of many millions of people throughout the rural Global South.

In a discussion paper prepared for the World Health Organization (WHO), Max Abbott says that "the gambling-related burden of harm appears to be of similar magnitude to harm attributed to major depressive disorder and alcohol misuse and dependence". He argues that the harm caused by gambling "is substantially higher than harm attributed to drug dependence disorder" and that "the burden is primarily due to financial impacts, damage to relationships and health, emotional/psychological distress and adverse impacts on work and education. This burden is disproportionately carried by disadvantaged and marginalised population sectors and contributes to health and social disparities".[11] The evidence base for this statement is slim as Abbott's paper was based on research in New Zealand and Australia only, and in my experience the damage caused by alcohol still far exceeds the damage caused by gambling, at least in rural regions in Africa and Asia. However, mobile phones and smartphones are bringing rural regions within far easier reach of the gambling industry. This offers a massive new and

naïve customer base, and I expect that Abbott's statement may well become true in the very near future.

When I first saw a concentration of problem gamblers in the Global South, the gambling industry was still almost entirely urban-based and urban-focused. It was in 1990, in Agadez, a Nigerien town at the edge of the Sahara Desert. Mostly young men (no women)[12] gathered in and around a gambling hall that was profitable enough for the French owner to be able to live in a villa that featured the town's only swimming pool. It was dark and gloomy inside, with lines of machines, and I remember this place so well because of the desperation on the faces of the men as they walked out. It reminded me of my years as a volunteer in a homeless shelter in Rotterdam, where roughly a fifth of the guests were homeless because of a gambling disorder.

The gambling hall in Agadez had a grave impact on problem gamblers, but the number of people affected was limited as this was a physical place, in the centre of town, and it was too far a walk for rural men to cause impulse gambling. For obvious reasons of customer density, gambling halls and betting shops commonly choose urban locations. There are lines and lines of betting shops in Kampala, for example, but I have never seen one in any of Uganda's small villages (except for those that are situated along a main road). The link between gambling and the proximity of gambling opportunities is well-evidenced, and means that the likelihood of lifetime gambling (i.e., of people ever having gambled), of impulse gambling and of building a gambling disorder is far lower in rural regions than in urban ones.[13] As an evaluator, I only very occasionally came across rural problem gamblers. A few men had used microfinance facilities to borrow money, or had forced their wives to borrow money, and had gambled it away. A blind woman in Malawi had received training and other support and, as a consequence, was able to grow crops in her garden – but she did not benefit from the harvest as her husband sold it and used it to feed his drinking and gambling habits. I remember these specific tragedies because they were unusual: in most rural programmes, gambling did not come up as an impediment for anything.

Gambling in both Asia and Africa is under-researched[14] and, because rural gambling is far less of a problem than urban gambling, rural gambling is rarely researched at all. However, two pieces of research confirm the gap between urban and rural gambling. First, a systematic review of gambling research in Africa found that lifetime gambling was prevalent in well over half (57%–73%) of youth (or sometimes of young men only, depending on the research) – but the only *rural* study in the review, conducted in Malawi, found only 16% prevalence of lifetime gambling.[15] Second, research among 150 rural and 150 peri-urban individuals in South Africa's KwaZulu-Natal (both groups with equal numbers of men and women) found that 75% of the people who had never gambled were rural people; that the vast majority of the people at risk were urban people (68% of people at low risk and 90% of people at moderate risk were urban); and that *all* the people assessed to be problem gamblers were urban people.[16]

When rural people do gamble, they are more likely than urban people to do this in card games, with dice, through coin spinning and by betting on animal fights (generally cockfights, but in Indonesia I saw a lively trade in betta, a fighting fish used in gambling). This is generally *personal* gambling: it is done in social groups, without a company that is taking a cut (though there might be people who receive payments, such as formal authorities if gambling is illegal, or the organiser or host).[17] Some research found that such forms of gambling can still be damaging for chronic losers[18] and that gambling debts may be a reason to migrate in order to escape from creditors.[19] Other research found no measurable effects on the long-term distribution of wealth,[20] or concluded that gambling was largely harmless as winners were expected to share their gains or face sanctions.[21] Either way, for a community as a whole, personal gambling is far less damaging than commercial, *impersonal* gambling, where the gambler plays against a slot machine or similarly faceless, company-owned gambling instrument. This is because gambling without a third party taking a cut is, in the long run and on average, budget-neutral for the individual gamblers (provided it does not require skills), and the money does not leave the community. I came across one piece of research that concluded that such gambling even had a *positive* effect on the local economy, as "the intense desire to play cards for money [...] is a powerful force motivating people to engage in cash-earning activities".[22]

However, rural gambling patterns are changing as rural regions adopt urban (and foreign) practices, also in relation to gambling. In the past, this was the case for card game gambling, which either introduced gambling to rural regions[23] or replaced indigenous forms of gambling.[24] In more recent years it is impersonal gambling, such as in the form of scratch cards and, most recently, Chinese slot machines that, at least in Ghana, became sufficiently affordable to present an attractive source of extra income for rural shop owners.[25]

Such scratch cards and slot machines bring gambling within reach of many rural people – but impulse gambling is limited to people who are visiting a shop that offers gambling options, or who live relatively close to such a shop. The more threatening trend started with radio stations capitalising on the spread of mobile phones by opening gambling lines that entice people to send charged text messages to answer easy questions, with the chance of winning money. These gambling options still exist, but in recent years they came to operate in parallel to gambling via smartphones, and gambling companies are now able to reach into people's own homes, also increasingly deep into rural regions, 24/7. In these rural regions, such companies find an audience that is large and, when it comes to gambling, naïve and inexperienced.

Their collective offer is enormous. I checked a Ugandan interface app for gamblers and found that it offers 11 companies to gamble through (Kagwirago, BetWay, BetPawa, EliteBet, BetLion, WorldStar, AbaBet, Betin, GrandVictoria, Next, and Betcity), each of which has a wide range of gambling options at any time. The industry is far larger still in Kenya.[26] The Kenyan (but partly Bulgarian-owned) SportPesa illustrates the size of the country's online betting business. It has signed sponsorships with the English Premier League football clubs Hull City, Arsenal, Southampton and Everton (though the latter has discontinued the agreement prematurely), and in the cases of Hull City and Everton the sponsorships were the best-paying sponsor agreements in the clubs' histories.[27] Even where online gambling is illegal, such as in India, "the constantly increasing technological advancement and internet penetration, coupled with easy accessibility to and affordability of smartphones, [gives rise to] a growing concern that more people will start gambling online, and subsequently, more people will have gambling-related problems".[28]

This virtual gambling trend is dangerous. Online gambling is new and it will take time for the negative effects to become sufficiently painful, widespread and visible to create awareness of its dangers. For now, gambling is widely seen as a legitimate activity,[29] and at least until the 2026–27 season[30] this will be reinforced by the world's most-watched and betted-on football league, the English Premier League, where all but three clubs[31] are sponsored by betting companies, thereby explicitly and very visibly linking the world's most popular sport with gambling.

Some people – and mostly adult men – prefer to place their bets in betting shops or social venues such as tea stalls, even if they have online gambling accounts, because of the social dimension,[32] and perhaps because betting companies recruit attractive women to staff their shops.[33] Such social settings have the drawback of peer pressure:

> See Apu [Apu means sister and refers to Sharifa Sultana, the interviewer], all of us do not play. But some of our bettor friends will still call us by name and say, "of course you do not play, because

you do not have money to play. Or you do not have any guts to play- oh Mamma's boy." This is ridiculously insulting and many of us even occasionally play to avoid such conversations.[34]

For many others, online gambling has the advantage of being discreet ("I love my privacy. This makes me gamble mostly online").[35] They gamble in isolation, and research in the arrangement of casino slot machines suggests that gambling in isolation impairs a gambler's control:

> Probable pathological gamblers preferred the cubicle or the isolated area arrangement. They did not like the against-the-wall and counter layouts because these were considered too visible and too susceptible to the distractions of the surroundings. Despite this preference, pathological gamblers unanimously believed that the isolated area promoted impaired control over their gambling habits.[36]

Online gambling pulls in people who are too young to be allowed entry into betting shops, even after considering that age restrictions are widely ignored in such shops.[37] Safety measures such as the requirement to click a button saying "I confirm that I am over 25 years old" do not pose a meaningful obstacle, and mobile money accounts can easily be borrowed from friends and older siblings. Online gambling draws in women, too. There were no women in that gambling hall in Agadez, back in 1990, or in many of Africa's and Asia's betting shops today, and in my evaluative work I have not encountered cases where livelihood interventions failed because of a woman's gambling. However, cultural norms matter less in the privacy of one's home and there is early (and as yet scant) evidence that the "numbers [of young women gambling] have recently started to increase for online betting".[38]

Importantly, online gambling is *always* available, easy to access and directly linked to mobile money wallets. These mobile wallets advance financial inclusion and have many useful roles to play, but for people who engage in online gambling, harm may outweigh these benefits. Research among a little over a thousand (1,040) digital borrowers in Kenya, of whom some 30% (304) were 'digital bettors', concluded that "bettors were shown to be more likely than non-bettors to be financially distressed, engage in welfare undermining coping strategies, and have inferior welfare outcomes".[39] Causality goes both ways: gambling can *cause* financial distress, and gambling can be a form of negative coping behaviour for people *in* financial distress.

A gambler's prospects

A growing betting industry does not *have* to be a problem. In a paper I quoted before, Max Abbott (at the time the Chairman of the International Gambling Think Tank) said that:

> There is little doubt that greater gambling availability has led to increased consumption and increased problems in many parts of the world. However, in both expanding and maturing markets [...] problem gambling rates can decline, sometimes markedly, rather than increase.[40]

Realising a reduction of gambling-caused problems in expanding gambling markets requires a combination of active restrictions and support systems. In rural regions in the Global South, these do not generally exist yet. There are few enforced restrictions in advertising and promotions (e.g., first bet free, every tenth bet free) on radio, mobile phones and smartphones. Enforced betting breaks and pop-up warnings barely exist, and limits on access time, bet size, numbers of bets placed, and speed of play are not commonly used. Instead, online companies introduced high-speed gambling options such as three-minute virtual football games, so that bettors do not have to wait 90 minutes of play time before knowing the result. Like drug use disorders, gambling disorders are, for now, a mostly urban problem, and harm reduction support for people with such disorders has yet to be developed in rural regions.[41]

The next chapter is about the digital divide. In many ways this divide reinforces existing inequalities, and narrowing it would bring ample benefits. However, narrowing this divide would also mean an expansion of the problems of online fraud and online gambling into ever-more remote rural regions.

Notes

1 Suri, T. and Jack, W. (9 December 2016) "The long-run poverty and gender impacts of mobile money", *Science*, volume 354, issue 6317, pages 1288–1292.

2 For example, Dawson, S. (18 January 2017) *Why does M-PESA lift Kenyans out of poverty?* Consultative Group to Assist the Poor. (The 'Consultative Group to Assist the Poor', or CGAP, is housed at the World Bank in Washington.)

3 Both are part of "Observation 3" on page 2 of Morawczynski, O. and Pickens, M. (August 2009) "Poor people using mobile financial services: observations on customer usage and impact from M-PESA", *CGAP Brief*, Consultative Group to Assist the Poor, 4 pages.

4 Bateman, M., Duvendack, M. and Loubere, N. (2019) "Is fin-tech the new panacea for poverty alleviation and local development? Contesting Suri and Jack's M-Pesa findings published in *Science*", *Review of African Political Economy*, volume 41, issue 161, pages 480–495. There were already some critical publications in the years before, but they were largely based on political (i.e., anti-capitalist) positioning rather than on empirical research. For a few examples, see pages 148–149 of Knorringa, P. *et al* (2016) "Frugal innovation and development: aides or adversaries?", *European Journal of Development Research*, volume 28, number 2, pages 143–153.

5 Gitonga, S. (30 July 2019) *Millions of Kenyans blacklisted by CRB over digital loan defaults*, Business Today.

6 CBK, KNBS, and FSD Kenya (April 2019) *2019 FinAccess household survey; access, usage, quality, impact*, Central Bank of Kenya, Kenya National Bureau of Statistics, and Financial Sector Deepening Kenya.

7 Kusimba, S. (2021) *Reimagining money; Kenya in the digital finance revolution*, Stanford University Press, with the quotation from page 47.

8 CBK, KNBS and FSD Kenya (April 2019) *2019 FinAccess household survey; access, usage, quality, impact*, Central Bank of Kenya, Kenya National Bureau of Statistics, and Financial Sector Deepening Kenya, with the percentages (with an extra decimal place) in Figure 5.2 on page 54.

9 For both statements in this sentence, see Arora, P. (2019) *The next billion users: digital life beyond the West,* Harvard University Press.

10 See Schiffauer, L. (2018) "Dangerous speculation; the appeal of pyramid schemes in rural Siberia", *Focaal – Journal of Global and Historical Anthropology*, volume 81, pages 58–71; Verdery, K. (1996) "Faith, hope and Caritas in the land of the pyramids, Romania, 1990–1994", chapter 7 in Verdery, K., *What was socialism, and what comes next?*, Princeton University Press.

11 Abbott, M. (2017) "The epidemiology and impact of gambling disorder and other gambling-related harm", *Discussion paper for the 2017 WHO Forum on alcohol, drugs and addictive behaviours*, World Health Organization, with the quotations on pages 1, 2, and 3.

12 Studies that focus on Western countries show roughly comparable gambling participation rates of men and women. For example, Simone McCarthy and her colleagues refer to four studies that came to this conclusion in McCarthy, S. *et al* (2019) "Women and gambling-related harm: a narrative literature review and implications for research, policy, and practice", *Harm Reduction Journal*, volume 16, article number 18, 11 pages. However, all Africa- and Asia-focused studies I read conclude that men gamble more than women; that *problem* gambling is more common among men than among women; and that women are often opposed to the gambling habits of their husbands and other male household members. For example, Bitanihirwe, B. K.Y. and Ssawanyana, D. (January 2021) "Gambling patterns and problem gambling among youth in Sub-Saharan Africa: a systematic review", *Journal of Gambling Studies*, volume 37, issue 3, pages 723–745; George, S. *et al* (2016) "A cross-sectional study of problem gambling and its correlates among college students in South India", *BJPsych* (i.e., British Journal of Psychiatry), volume 2, issue 3, pages 199–203; and Nooteboom, G. (2015) *Forgotten people: poverty, risk and social security in Indonesia; the case of the Madurese*, Brill, see page 177.

13 Tagoe, V.N.K, Yendork, J.S. and Asante, K.O. (Spring 2018) "Gambling among youth in contemporary Ghana: understanding, initiation and perceived benefits", *Africa Today*, volume 64, issue 3, pages 52–69; Pearce, J. *et al* (October 2008) "A national study of neighbourhood access to gambling opportunities and individual gambling behaviour", *Journal of Epidemiology and Community Health*, volume 62, issue 10, pages 862–868; and Serge, S. *et al* (2008) "Links between casino proximity and gambling participation, expenditure, and pathology", *Psychology of Addictive Behaviors*, volume 22, issue 2, pages 295–301.

14 In Asia, gambling research tends to focus on specific locations (e.g., Hong Kong, Macao and Singapore). For example, a 2017 review of gambling research in India, where gambling is largely illegal but voluminous (with an annual turnover of US$42 billion, according to one estimation), concludes that "To the best of our knowledge, there have only been three studies of gambling from India". George, S., Velleman, R. and Nadkarni, A. (2017) "Gambling in India: past, present and future", *Asian Journal of Psychiatry*, volume 26, pages 39–43, with the quotation on page 41. A passing reference to this statement, made in 2020 by the same lead author, suggests that nothing new had appeared in the last few years – see George, S., Fenn, J. and Robonderdeep, K. (2020) "An overview of gambling in India", *Global Journal of Medical, Pharmaceutical and Biomedical Update*, volume 15, number 4, pages 1–4. The US$42 billion estimate is from 2013 and comes from George, S., Velleman, R. and Weobong, B. (July 2020) "Should gambling be legalised in India?", *Indian Journal of Psychological Medicine*, volume 43, issue 2, pages 163–167, with the figure on page 164 (with a reference to a 2013 publication in which I was unable to verify the figure). In Sub-Saharan Africa, a 2021 systematic review of publications on youth gambling (and note that most gambling research focuses on youth) found

only 13 publications that reported on data-driven research in peer-reviewed journals in all of Sub-Saharan Africa: Bitanihirwe, B.K.Y. and Ssawanyana, D. (January 2021) "Gambling patterns and problem gambling among youth in Sub-Saharan Africa: a systematic review", *Journal of Gambling Studies*, volume 37, issue 3, pages 723–745.

15 *Ibid*. The Malawi study is Muchimba, M. *et al* (2013) "Behavioral disinhibition and sexual risk behavior among adolescents and young adults in Malawi", *PLoS ONE*, volume 8, issue 9, 6 pages.

16 Dellis, A. *et al* (2013) "Gambling participation and problem gambling severity among rural and peri-urban poor South African adults in KwaZulu-Natal", *Journal of Gambling Studies*, volume 29, number 3, pages 417–433, with the percentages in Figure 1 and the sample numbers in Table 1. The remaining category, "no risk", was split 50:50.

17 Organised group gambling does have systematic 'leakage'. In cockfights in Flores in Indonesia, "Winners double their money, [minus] ten per cent for the house, five per cent for the landowner, and five per cent for security – army and police". Curnow, J. (February 2012) "Gambling in Flores, Indonesia: not such a risky business", *The Australian Journal of Anthropology*, volume 23, pages 101–116, with the quotation from page 109.

18 "Some people are chronic losers, whereas others seem to be successful most of the time [and] some informants suggested that a few chronic losers were also those with low incomes." Grossman, L.S. (1984) *Peasants, subsistence ecology, and development in the highlands of Papua New Guinea*, Princeton University Press, with the quotation on page 204.

19 "Many of the Madurese migrants looking for work in East Kalimantan [...] gave gambling debts as a reason for migration." Nooteboom, G. (2015) *Forgotten people: poverty, risk and social security in Indonesia; the case of the Madurese*, Brill, with the quotation from page 177.

20 "The overall redistribution of gambling money in the long run has not created perceptible differences in wealth or privilege within the village." Hayano, D.M. (September 1989) "Like eating money: card gambling in a Papua New Guinea highlands village", *Journal of Gambling Behavior*, volume 5, pages 231–245, with the quotation from page 241.

21 "A big win [...] is socialised cash. [The winners] can expect to receive loan requests. Quietly stashing away money is seen as stingy, and while the cash may be hidden from the living, it places the individual in danger of supernatural sanctions." Curnow, J. (February 2012) "Gambling in Flores, Indonesia: not such a risky business", *The Australian Journal of Anthropology*, volume 23, pages 101–116, with the quotation from page 112.

22 Grossman, L.S. (1984) *Peasants, subsistence ecology, and development in the highlands of Papua New Guinea*, Princeton University Press, with the quotation on page 204.

23 David Hayano describes how card game gambling moved from colonisers to the urban areas of Papua New Guinea, and from there, via migrant workers, to the highlands. Hayano, D.M. (September 1989) "Like eating money: card gambling in a Papua New Guinea highlands village", *Journal of Gambling Behavior*, volume 5, pages 231–245.

24 In Ngada in Indonesia, "corn kernels were a popular device for gambling prior to the introduction of playing cards. A croupier would scoop up a handful of loose corn kernels and deal out four piles. Punters would then gamble on how many kernels were left in the croupier's hand. Animals, knives or anything else of value were wagered". Curnow, J. (February 2012) "Gambling in Flores, Indonesia: not such a risky business", *The Australian Journal of Anthropology*, volume 23, pages 101–116, with the quotation from page 104.

25 Hayk, A. and Sailer, U. (May 2020) "Cosmopolitan encounters provoke a change in habits: how Chinese slot machines affect rural life in Ghana", *Geoforum*, volume 111, pages 39–47, discusses the effects of such slot machines, which Chinese operators installed in parts of rural Ghana in the summer of 2016.

26 The "total financial turnover [of the top 12 Kenyan websites only] is estimated at 250 billion [Kenyan shillings, which, at the time, equated to some US$2.4 billion], more than the national recurrent budget", according to Amutabi, M.N. (February 2018) "Gambling addiction and threat to development in Kenya: assessing the risks and problems of gamblers in changing society", *Journal of African Interdisciplinary Studies*, volume 2, number 2, pages 121–133, with this quotation and an overview of the companies behind these 12 websites on page 122. The national recurrent budget is the part of the government budget that covers ongoing expenses (as opposed to capital expenses).

27 For Everton, see Conneller, P. (17 February 2020) *EPL team Everton aborts SportPesa sponsorship deal two years early*, Casino.org. Everton discontinued the contract following mounting criticism about SportPesa's role in the 'gamblification' of football. For Hull City, see SportPesa (undated) *SportPesa joins up with the tigers!*, SportPesa, which quotes the Hull City Commercial Managers as saying that "The deal is by far the largest in our history and a sign of our ambition to grow our Club and become a household name in our sport globally".

28 George, S., Fenn, J. and Robonderdeep, K. (2020) "An overview of gambling in India", *Global Journal of Medical, Pharmaceutical and Biomedical Update*, volume 15, number 4, pages 1–4, with the quotation on page 2.

29 Many of the papers referenced in this section come to this conclusion, in both Africa and Asia. See several of the publications reviewed in Bitanihirwe, B.K.Y. and Ssawanyana, D. (January 2021) "Gambling patterns and problem gambling among youth in Sub-Saharan Africa: a systematic review", *Journal of Gambling Studies*, volume 37, issue 3, pages 723–745; Curnow, J. (February 2012) "Gambling in Flores, Indonesia: not such a risky business", *The Australian Journal of Anthropology*, volume 23, pages 101–116; and Grossman, L.S. (1984) *Peasants, subsistence ecology, and development in the highlands of Papua New Guinea*, Princeton University Press.

30 In April 2023, the Premier League clubs agreed to stop featuring gambling adverts on the front of their shirts, as of the 2026–27 football season. See SIG (14 April 2023) *Premier League confirms gambling sponsorship ban*, Sport Industry Group

31 In the 2020–21 season, only Sheffield United, Liverpool and Chelsea were not in a sponsorship agreement with a betting company. Mancunian Matters (11 January 2021) *The evolution of betting sponsors in the English Premier League*, Mancunian Matters.

32 In betting shops: Adebisi, T. *et al* (March 2021) "Gambling in transition: assessing youth narratives of gambling in Nigeria", *Journal of Gambling Studies*, volume 37, issue 1, pages 59–82, with the point made on page 75. In tea shops: Sultana, S., Mozumber, M.H. and Ahmed, S.I. (May 2021) "Chasing luck: data-driven prediction, faith, hunch, and cultural norms in rural betting practices", *CHI Conference on Human Factors in Computing Systems (CHI '21)*, Yokohama, Japan.

33 I base this on my own observations and an interview with a young East African woman, with whom I discussed livelihoods options, and who matter-of-factly mentioned that there was a group of opportunities, such as running a street-side tea shop or working in a betting shop, that she did not qualify for as she was insufficiently attractive.

34 Sultana, S., Mozumber, M.H. and Ahmed, S.I. (May 2021) "Chasing luck: data-driven prediction, faith, hunch, and cultural norms in rural betting practices", *CHI Conference on Human Factors in Computing Systems (CHI '21)*, Yokohama, Japan, with the quotation, from a 20-year-old male student, on page 11.

35 The quotation is part of a longer quotation from a survey respondent, quoted in Adebisi, T. *et al* (March 2021) "Gambling in transition: assessing youth narratives of gambling in Nigeria", *Journal of Gambling Studies*, volume 37, issue 1, pages 59–82, with the quotation from Table 4 on page 70.

36 Ladouceur, R. *et al* (2015) "Impact of the format, arrangement and availability of electronic gaming machines outside casinos on gambling", *International Gambling Studies*, volume 5, number 2, pages 139–154, with the quotation on page 144.

37 A BBC documentary shows just how easy it is for underage gamblers to place bets, both online and in betting shops. BBC News Africa (2019) *Gamblers like me: the dark side of sports betting*, BBC.

38 Mustapha, S.A. and Enilolobo, O.S. (December 2019) "Effects of gambling on the welfare of Nigerian youths: a case study of Lagos State", *Journal of Gambling Issues*, volume 43, pages 29–44, with the quotation from page 36.

39 Chamboko, R. and Guvuriro, S. (February 2021) "The role of betting on digital credit repayment, coping mechanisms and welfare outcomes: evidence from Kenya", *International Journal of Financial Studies*, volume 9, number 10, 12 pages, with the quotation from page 10.

40 Abbott, M. (2017) "The epidemiology and impact of gambling disorder and other gambling-related harm", *Discussion paper for the 2017 WHO Forum on alcohol, drugs and addictive behaviours*, World Health Organization, with the quotation on page 4.

41 I base this last statement on work I did in 2013–2016, for a programme titled 'Asia Action on Harm Reduction'. This programme focused on organisations supporting (and often owned and managed by) drug users. These organisations, and the organisations in their wider networks, all focused almost exclusively on urban centres (and occasionally key transport hubs). The statement makes two assumptions. First, Asia Action was focused on

drug users, not problem gamblers, and I assume a parallel because both are predominantly urban phenomena in which harm reduction work takes the shape of small-scale interventions. Second, my work for Asia Action took place in Asian countries only (Cambodia, China, India, Indonesia, Malaysia and Vietnam) and I merely assume that the situation is similar in Africa on the basis of never having come across support facilities that did exist in rural Africa but not in rural Asia.

12 The digital divide and its effects

Remote rural people, and disadvantaged groups among them, could often potentially benefit more from ICT opportunities than anybody else, because travel costs and other obstacles mean that more traditional methods of communication are difficult and limited. However, the actual opportunities ICT presents them with are more modest than those of other groups, as they are on the wrong side of the aforementioned – but not yet clarified – 'digital divide'.

The digital divide is the gulf between those who easily access and use modern ICT and those who do not. The digital divide aligns with other forms of marginalisation and vulnerabilities (see Box 12.1) and therefore reinforces existing inequalities. An older illiterate rural woman in the Global South, for example, is on the wrong side of at least five divides: the education divide, the urban–rural divide, the gender divide, the age divide and the North–South divide. Other groups that "disproportionately remain offline [are] those with disabilities, indigenous populations and some people living in the world's poorest places".[1] These divides are not independent but reinforce each other. For

DOI: 10.4324/9781003451716-15

example, "women are less likely than men to make use of the internet in most countries, but are *more* underrepresented online in developing countries than in developed countries, and are *especially* under-represented in [the least developed countries]".[2] Rural girls are not merely less likely than urban boys to be ICT literate and know about the opportunities ICT could potentially provide. They are *also* less likely to have a phone or to be at liberty to use a shared phone *and*, if they do have one, they are less likely to have coverage, credit, and their parents' permission to use it for other purposes than to confirm their safe arrival at school. They might have less time as well, if they combine education with household chores. All groups fall into smaller sub-groups of people with characteristics that may reduce or enhance their chances to take part in a digitalised world. A women-focused ICT training course on Zanzibar attracted almost exclusively unmarried women without children, for example, because only they could spare the time.[3]

The divide is both a general phenomenon and a phenomenon that exists in agriculture in particular, where the divide manifests itself in terms of both adoption and usefulness. This is the case *within* countries, for the reasons just mentioned, but also *across* countries:

> ICT adoption levels of richer countries are much higher than those of poorer countries, and [...] returns to ICT in agricultural production in richer countries are about two times higher than in poorer countries. A plausible explanation for the poorer countries' relatively low productivity elasticity of ICT is the lack of important complementary factors, such as better electricity and transportation infrastructure, productive human capital (stemming from higher levels of education), and business models that have been transformed to deal with the information age.[4]

Box 12.1 The digital divide and climate change

Many of the people who are on the wrong side of the digital divide are also likely to be disproportionately affected by the effects of climate change. There are two reasons for this.

1   The remote rural areas that do not have phone and internet coverage are also often areas that are vulnerable to hazards that are aggravated by climate change, such as wind erosion, drought and heat.

2 The groups with least access to ICT – such as remote rural women, older people and illiterate people – often don't have easy access to other channels of information and advice either. They are therefore less likely to gain the knowledge needed to shield their plots from erosion, to harvest scarce rainwater or to minimise evaporation. This reduces their income-earning potential and increases their exposure to shocks.

Both climate change and the digital divide consolidate and deepen the inequalities between remotely located and particularly poor people, and their less remote and wealthier peers.

Earlier pages covered many examples of ICT applications that aim to incorporate small-scale farmers into corporate value chains, in a drive that is eagerly supported by donors, the World Bank and some other multilateral agencies, and the private sector. It also mentioned examples of 'frugal innovations' that provide core functionality at very low (financial)[5] cost, such as M-Pesa's mobile money, Jaguza's ear tags and Widim Pump's remote irrigation management system; and free hacks, such as the 'ring once to say you arrived' trick and illegal downloads. Notwithstanding these and many other examples of accessible and free or affordable ICT opportunities, the *general* rules are probably that (1) anything that is not produced locally is more expensive if you are poor and live in a remote rural region; and that (2) irrespective of costs, remote rural people are the least likely to have access to ICT opportunities. This is because they are the least connected people in the least connected parts of the world, and they are difficult and commercially unattractive target groups on which to focus product development. There are three reasons for this:

1 **Unfamiliarity with poor rural people's ICT needs.** People specialised in ICT product development rarely have poor rural backgrounds themselves, and therefore do not have their needs in mind when developing products. Even if they try, the results may not be very good because of their lack of real understanding of the needs and possibilities in poor rural regions. In principle, developers could work closely with their target audience. Indeed, a 2019 review of reviews concluded that "there is a need to shift from developing technologies *for* users to designing and developing applications *with* users to enhance collective problem definition and gain more insight of the context that can significantly

affect the outcome of the intervention".[6] In this spirit, Richard Heeks developed the *ladder of inclusive innovation* and argued that ICT innovations stand more chance of generating inclusive development results if they are designed following the higher-level principles of inclusivity in the product development stage (see Box 12.2). The review's observation is valid and Heeks's ladder is a useful target-setting benchmark, but inclusivity comes at a cost: the higher the level of inclusivity, the more demanding and time-consuming the innovation process. Not many ICT companies are willing to make that investment.

2   **A lack of commercial appeal.** It is generally commercially unattractive to develop a new and rural-focused ICT product or service that is 'frugal': developed with the intention of achieving extreme cost savings while maintaining core functionality, in order to reach poor rural people.[7] The profit prospects are unlikely to be good, as potential customers are dispersed, often ICT illiterate and with few means to be made aware of new opportunities, and they have relatively little purchasing power. This is why many of the frugal rural ICT products and services that do exist have been developed with the support of government or donor funding... and even *those* do not generally stretch very deep into rural regions and often do not reach the poorest segments of society. For example, Jaguza's ear tags and Widim Pump's remote irrigation management system are only useful for people with cattle or land – and poor rural people often have neither.

3   **Poor ICT infrastructure.** Both inclusive and frugal innovations are hindered by the absence or unreliability of ICT infrastructure in remote areas. Such hindrances can be overcome, with solar energy innovations (which include several frugal innovations) and by working around connectivity challenges (e.g., many of the apps mentioned in this book have some offline functionality). However, this does add to the time and attention required for the development of inclusive and/or frugal innovations, and therefore further reduces their commercial appeal.

For decades now, development-focused NGOs have sought to address these digital divides. They invented some of the products mentioned in this book, invested in digital education and awareness-raising, and sometimes used the higher levels of Heeks's ladder of inclusive innovation when doing so. Their role has been helpful, sometimes, but NGOs are not generally specialist ICT developers. Many of their endeavours have failed, and a statement that Lucas Joppa made in reference to the development of ICT solutions in the field of nature conservation applies to NGOs' ICT solutions in other fields of rural development as well: "The current general approach is a patchwork of one-off projects and partnerships".[8]

Box 12.2 Heeks's 'ladder of inclusive innovation'

Level 1: Intention. The intention of an ICT product or service is to be of use to a group that is poor and often overlooked.

Level 2: Consumption. The innovative product is affordable to, visible for and actually adopted and used by that group.

Level 3: Impact. The innovation has a positive impact on the livelihoods or well-being of the group. Existing inequalities reduce because the positive impact on the disadvantaged group is larger than the impact on other groups.

Level 4: Process. Members of the group are involved in the design, development, production and distribution of a specific innovation. Being involved could mean all sorts of things, from simply being consulted to having an actual say in or even control over an innovation.

Level 5: Structure. Innovations are created within a structure that is itself inclusive. The involvement of members of the normally excluded group is part of standard procedures, and not merely a temporary convenience to the innovators.

Source: This is paraphrased and summarised from Heeks, R. *et al* (2013) "Inclusive innovation: definition, conceptualisation and future research priorities", *Working Paper Series number 53*, Centre for Development Informatics. I thank Richard Heeks for his feedback on a draft version of this box.

In my experience, private sector companies – often start-ups – that receive financing and other support from impact investors and ODA-financed investment funds stand a better chance of success. Most of the examples given in this book come from social and commercial companies that, at least in the initial stages, received financial support and/or concessional financing. Their survival chances are better than those of NGO projects, in part because these companies have greater expertise and in part because their solutions are, ultimately, profit-driven, and by design less likely to require ongoing support after that initial insertion of ODA funding.[9] They are also more experienced in marketing. This is vital because, as you will remember from Chapter 8, people won't use products they do not know exist, *and* they need to feel confident of its usefulness and user-friendliness.

Neither NGOs nor tech companies are in a position to arrange the necessary infrastructure – internet connectivity and electricity supply – that ICT products and services require (though this is changing in the case of electricity, because of the expansion of off-grid options).[10] Building and maintaining such infrastructure is generally the prerogative of specialist infrastructure companies. These companies will not pursue full coverage as they make their investments largely on the basis of "potential market size [...]; per capita income; installation and maintenance cost factors related to accessibility (elevation, slope, distance from a main road, and distance from the nearest large city); and national competition policy".[11] The general rule is that more competition sparks better coverage, and the just-quoted paper shows this causal link to be very strong. With or without competition, these companies are unlikely to cover the 'last mile connectivity' in the harder-to-access regions with more dispersed populations (such as the forest populations discussed before) unless they are incentivised to do so. Three publications from Jenny Aker and her colleagues claim that this 'last mile' problem is often resolved because "'last mile' regulatory policies designed to connect marginalized individuals are now commonly included in the licenses granted to mobile operators".

It is not clear what evidence Aker and her colleagues base this on,[12] but they may well be right as infrastructure costs are coming down and most governments have either achieved near-comprehensive coverage or are moving towards that point. Between 2015 and 2020, global 4G network coverage doubled, and the International Telecommunication Union predicted that "globally, almost 84 per cent of the population will be covered by a 4G network at the end of 2020". In Asia and the Pacific, coverage is even higher (over 94%). In Africa it is significantly lower (44%, plus 33% on 3G and 11% on 2G), but Africa is catching up and "In 2020, Africa achieved 21 per cent growth in 4G rollout, while growth was negligible in all the other regions".[13]

Not everybody who has coverage and a smartphone uses the internet, but an increasing number does, if mostly for recreational rather than development-related purposes (see Box 12.3). After many years of uninterrupted growth, the world crossed the 50% mark of the world's people using the internet in 2019 (it was 54% by year end, and then progressed to 66% in 2022),[14] and for at least three reasons this proportion will continue to increase year-on-year:

1 Coverage will continue to grow. The fact that Africa was the only continent on which it grew in 2020 was the consequence of the pandemic, rather than a trend reversal.

2 Costs of internet usage are coming down, though they are still lower in the West than in the Global South, and the average costs in the Global South are still well above the UN target of less than 2% of per capita gross national income, or GNI, with the highest percentage (6.4%) in Africa.[15]

3 One of the most important digital divides is age,[16] and the unconnected older people will gradually be replaced with better-connected younger ones, who will not stop being online just because they get older.

For a number of reasons, digital divides will continue to exist. Without donor and government funding, companies will naturally focus most of their product and service development on clients who have significant purchasing power and are easy to reach, and these are wealthy clients who live in countries with many such clients. Rural regions are increasingly covered by 3G and 4G, but the divide persists as cities are moving to 5G. Young, educated urban groups will continue to gain far more sophisticated ICT skills than older, uneducated rural groups. And ICT adoption by low- and middle-income countries will continue to lag behind high-income and economically more advanced and diversified countries.

Box 12.3 Manhunt and Bulletstorm

Demand for online ICT applications does not necessarily, or even usually, lead to socio-economic development, empowerment of women or better and more secure livelihoods. I have rarely seen women in rural internet cafés in Egypt, for example, and instead their clientele consists almost exclusively of boys and young men who use the computers mostly to play violent games.

In a 2016 paper for the *International Journal of Communication*, Payal Arora says that "primarily, people go online to romance, game, be entertained, consume media, view pornography, and share their personal thoughts and feelings".[17] The international development community by and large ignores this fact, and has a near-exclusive focus on productive, development-related uses of ICT applications. This is the consequence of the combination of a blind spot and a sense of discomfort with the notion that poor people might enjoy entertainment.[18] NGOs and companies reinforce this donor blind spot. With the exception of 'learning-while-playing' messages in education-related reports, I do not think I have ever seen a donor report in which a company or NGO covers the recreational use of whatever product or service it sold or provided to its 'beneficiaries' (i.e., customers).

However, these divides may prove to be less fundamental than the original divide between people with and without access to mobile phones and internet, and between people with and without basic ICT skills. M-Pesa's ability to function on 2G, and many of the agricultural apps that are designed to be used offline or with very little bandwidth are proof that the argument that "all the opportunities that the digital revolution represents are premised on super-fast, reliable and affordable connectivity"[19] is just not true. In addition, the evidence of the benefits that ICT applications potentially have for the poorest segments in the rural Global South is by now strong enough for donors to continue their financial support to efforts in this field for the foreseeable future (all the more so because these benefits are, as discussed above, significantly exaggerated in the minds of donors). By means of illustration: I often support donors and NGOs with the development of their Theories of Change, and in recent years these Theories of Change almost invariably mention a particular focus on ICT solutions.[20] Lastly, while this book's prologue gave examples of ICT supporting learning in rural schools, it did not mention that the use of ICT applications in these schools, in and of itself and irrespective of the subject matter being taught, helps break down the digital divide in the next generation.

So it seems to me that the digital divide of the future will be less dramatic than the digital divide of the past has been. This is, however, a very tentative prediction, as I base it on past and current realities while ICT has a strong 'black swan' propensity: it comes with unexpected innovations and shifts in usage that rather suddenly appear and that sometimes change everything we thought we knew to be true.

Notes

1 This quotation is from UN News (December 2018) *Internet milestone reached, as more than 50 per cent go online: United Nations telecoms agency*, United Nations.

2 ITU (2017) *Measuring the information society report 2017*, volume 1, International Telecommunication Union, with the quotation from page iv (emphasis added).

3 This is just one of many examples. It comes from McCarrick, H. and Kleine, D. (2017) "Digital inclusion, female entrepreneurship, and the

production of neoliberal subjects – views from Chile and Tanzania", chapter 4 of Graham, M., editor, *Digital economies at global margins*, The MIT Press.

4 Lio, M. and Liu, M.C. (2006) "ICT and agricultural productivity: evidence from cross-country data", *Agricultural Economics*, volume 34, number 3, pages 221–228, with the quotation from page 227.

5 Sometimes there are parallel streams of literature that are very alike but never meet. Ecomodernist literature is about frugal innovation in environmental footprint terms, but it is not part of, and rarely mentioned by, the frugal innovation stream of literature, which focuses almost exclusively on frugality in financial cost terms.

6 Lwoga, E.T. and Sangeda, R.Z. (2019) "ICTs and development in developing countries: a systematic review of reviews", *The Electronic Journal of Information Systems in Developing Countries*, volume 85, issue 1, 17 pages, with the quotation from page 10; emphasis added. This statement seems to be mostly based on only one of the reviews Lwoga and Sangeda reviewed, which is Zewge, A. and Dittrich, Y. (2017) "Systematic mapping study of information technology for development in agriculture (the case of developing countries)", *The Electronic Journal of Information Systems in Developing Countries*, volume 82, issue 2, 25 pages.

7 For the conceptual differences between inclusive innovation and frugal innovation, see Onsongo, E. and Knorringa, P. (June 2020) "Comparing frugality and inclusion in innovation for development: logic, process and outcomes", *CFIA Working Paper Series number 7*, Centre for Frugal Innovation in Africa.

8 Joppa, L.N. (2015) "Technology for nature conservation: an industry perspective", *Ambio*, volume 44, supplement 4, pages S522–S526, with the quotation from page S523.

9 This comparison is based solely on my own experience, and it is specific to the development of ICT solutions. In other fields, NGOs outperform private sector companies. For research that compares their respective performances in the field of democratisation efforts and concludes that NGOs outperform private sector companies, for example, see Dunton, C. and Hasler, J. (2021) "Opening the black box of international aid: understanding delivery actors and democratization", *International Politics*, volume 58, pages 792–815.

10 The uptake and impact of the previously mentioned off-grid Mkopa services is covered in Mutongwa, S.M. and Abeka, S. (2020) "Implementation of MKopa solar services for poverty eradication", *Journal of Scientific and Engineering Research*, volume 7, number 1, pages 76–87. There are a number of other pay-as-you-go off-grid energy providers (e.g., SHS, Zola), but Ben Leo and his colleagues concluded from a survey in 12 African countries that people with solar kits do not feel these kits fulfil their energy needs and they still desire grid electricity (the same principle applies to wind turbine kits). See Leo, B., Kalow, J. and Moss, T. (February 2018) "What can we learn about energy access and demand from mobile-phone surveys? Nine findings from twelve African countries", *CGD Policy Paper 120*, Center for Global Development, with the issue covered on pages 14–15.

11 The most thorough research on this has been done in Sub-Saharan Africa and is reported in Buys, P. *et al* (September 2009) "Determinants of a digital divide in Sub-Saharan Africa: a spatial econometric analysis of cell phone

coverage", *World Development*, volume 37, issue 9, pages 1494–1505, with the quotation from page 1502.

12 This quotation is from Aker, J.C. and Blumenstock, J.E. (November 2014, print version July 2015) "The economic impacts of new technologies in Africa", *The Oxford Handbook of Africa and Economics, volume 2: policies and practices*, with the quotation from section 19.2.1. The same text is used in Aker, J.C., Ghosh, I. and Burrell, J. (2016) "The promise (and pitfalls) of ICT for agriculture initiatives", *Agricultural Economics*, volume 47, supplement, pages 35–48, on page 38. Both papers refer to Aker, J.C. and Mbiti, I.M. (2010) "Mobile phones and economic development in Africa", *Journal of Economic Perspectives*, volume 24, issue 3, pages 207–232, which says nothing about public–private partnerships or contracts that are conditional on last mile coverage.

13 Both quotations and all figures are from ITU (undated) *Measuring digital development; facts and figures 2020*, International Telecommunication Union, section on mobile network coverage. Side note: ecomodernists could use the move from 2G to higher-level generations as an example of technological innovation leading to both better products and lower carbon footprints, as later generations are far more energy efficient than earlier generations, *and* some of the tech giants that most depend on 4G and 5G (e.g., Apple, Google, Microsoft) have moved to 100% use of renewable energy sources. See Spotlight 5.2 on pages 186–187 of World Bank (2021) *World development report: data for better lives*, A World Bank Group Flagship Report.

14 For both the growth trend and the 50% crossing in 2019, see the statistics page of the ITU website (itu.int/en/itu-d/statistics/pages/stat/default.aspx), accessed on 7 September 2023.

15 This is the target of the ITU/UNESCO Broadband Commission for Sustainable Development. See the ITU's web page on "affordability": itu.int/en/mediacentre/backgrounders/Pages/affordability.aspx. See also ITU (undated) *The affordability of ICT services 2020; policy brief*, International Telecommunication Union.

16 This is a very consistent finding. To give but one of many examples, see Zhu, Z., Ma, W. and Leng, C. (2020) "ICT adoption, individual income and psychological health of rural farmers in China", *Applied Research on Quality of Life*, 21 pages, with the fact presented on pages 9–10 and in Table 2.

17 Arora, P. (2016) "The bottom of the data pyramid: big data and the Global South", *International Journal of Communication*, volume 10, pages 1681–1699, with the quotation from page 1692.

18 Arora, P. (2019) *The next billion users: digital life beyond the West*, Harvard University Press.

19 This type of argument is presented regularly, by think tanks and multilateral organisations. This particular quotation comes from page 5 of Akileswaran, K. and Hutchinson, G. (July 2019) *Adapting to the 4IR: Africa's development in the age of automation*, Tony Blair Institute for Global Change.

20 In the first half of 2021, for example, I supported a European donor with the development of three Theories of Change. According to the one-page summaries of these Theories of Change, one aimed to achieve "digital infrastructural improvements"; the second included the statement that "we support and encourage the replication and scaling up of successful (digital) innovations"; and the third one is, in part, "focused on digital opportunities".

13 Misinformation

In 1990, when I first visited Africa, I went hunting with a group of Mbuti in DRC – then called Zaire. I was in awe. At every turn they saw, felt and smelled things that I did not understand the significance of. Since that day of hunting, I have often been impressed by the intricate knowledge indigenous populations have of their environments.

However, I have also come across many bits of 'knowledge' that may have cultural significance but otherwise seem incorrect. Some of it is innocent. In Bosnia and Herzegovina I was warned against leaving the house with wet hair, as this would cause my mouth to move from the centre of my face to the left. Other 'insights' are not so innocuous. The proximity of a menstruating woman does not in fact affect the quality of the harvest, and the Yemeni Muhamasheen (meaning 'slaves') should not be seen as rights-less people who can be exploited at will. Some misconceptions have caused instant death. When I visited Pakistani Kashmir after the 2005 earthquake I learned that many children had died in school because their teachers had told them to remain motionless at their desks, lest their movements should aggravate the earthquake.

The teachers had not made this up. They had received misinformation. Misinformation is of all times and places. It comes from inside communities and from the outside. It is not limited to or predominantly from uneducated people, and it is often conveyed with good intentions.

The discourse around genetically modified organisms (GMO) serves as an illustration. Most people advocating against GMO mean well, but many of their arguments are based on misinformation. Because the opposition to GMO is firmly in the mainstream of the discourse, people are at liberty to convey their anti-GMO messages without the need to

DOI: 10.4324/9781003451716-16

provide evidence. Indeed such evidence often does not exist.[1] Some people reject these 'Frankenstein foods' because it is not up to humankind to tamper with God's creations.[2] Others – including supporters of the degrowth movement[3] and many advocacy NGOs[4] – oppose genetically modified crops because they see risks related to consumer safety; they fear that genetically engineered species will spread into nature; or they foresee the development of super-pests and super-weeds that are trained by GM crops' toxins (i.e., most pests and weeds would die, but the ones that survive and spread would be resistant to these toxins). Some anti-GMO advocates point to the theory that GMO varieties drive poor farmers into destitution and towards suicide (see Box 13.1). There are also completely made-up arguments. For example, a UK-based NGO working in Zambia suggested that genetically modified maize might introduce a retrovirus akin to HIV; and several NGOs have argued that GM seeds have a 'terminator gene' that renders the seeds sterile[5] (note the contradiction with the threat of engineered species spreading into nature).

The only argument for which there is persuasive scientific evidence is the one related to the development of super-weeds and super-pests. Most of their development is not GMO specific and is instead caused by the very invention of farming,[6] but GMO has played a role as well – for example, pest resistance to the *Bt* bacteria, genetic material of which is incorporated in some GM seeds, is on the rise.[7] This argument does not disqualify GM crops in their totality. It merely questions the wisdom of developing GM crops that may lead to a natural selection process that eventually turns weeds and bugs into super-weeds and super-bugs. I cannot think of a similar drawback in relation to GM crops that are, for example, drought resistant or packed with nutrients.

The net effect of at least some GM crops on poverty and food security has, for now, been positive,[8] even though GM crops are mostly used for cotton and animal fodder, and only rarely for food production. In all likelihood, the positive effects would have been far stronger if misinformation had not been so effectively used in advocacy efforts that held back GMO research investments in crops that are specifically designed to suit the localised needs of small farmers or serve the purpose of adapting agriculture to climate change. In the few cases where useful GMO food crops *were* developed (e.g., potato, sweet potato, cassava, sorghum), widely spread misinformation has resulted in such conservative regulation, in most countries in the Global South, that it renders the cultivation of GM crops all but impossible.[9]

Box 13.1 GM crops and pricing traps

A common accusation is that genetically modified varieties drive poor farmers into destitution because subsidies and free trials inspire them to adopt GM crops and then, when they have adapted their farming practice and have become dependent on GM seeds, they become indebted when the patent-holders increase their seed prices. A frequently made claim is that this leads to suicide, particularly among Indian cotton farmers (250,000 of them to date, by a commonly given but never substantiated estimate).[10]

The prevalence of suicides among Indian farmers is undisputed, but it is not proven that these are 'GMO suicides' caused by 'suicide seeds'. An alternative explanation for the very high numbers of Indian farmer suicides is financial distress caused by India's banking reform, which led to higher prices and lower availability of agricultural credit.[11]

Moreover, even if the claims were true, they would not be the consequence of an *inherent* problem of GM seeds, but the result of marketing and pricing techniques that are made possible by "a form of radical monopoly of the food system" that, degrowth scholars say, "represents a threat to humanity".[12] In my evaluation work I have seen similar pricing dynamics for fertilisers (both deliberately, by commercial companies, and inadvertently, by changes in subsidy policies) and in private education, where parents are lured into schools that charge very little, only to find that every subsequent year is a little costlier.[13] The best-known price trap catastrophe was related to formula milk, where mothers were encouraged to 'try' formula milk until their own milk had dried up, creating dependency and excess infant mortality. Such damaging pricing techniques exist for GM seeds as well, but there are also examples of companies licensing their technologies on a royalty-free basis[14] and of local companies producing pirated seeds.[15]

The formula milk catastrophe led to marketing codes, which prevented much further suffering and death. Such marketing codes, and government-controlled pricing, could resolve the practice of pricing traps in GM seeds (and in fertilisers and in education) as well.

Whilst misinformation is of all times and most GMO-related misinformation was first conveyed well before the advent of online and social media, it was the internet that allowed misinformation to mushroom into massive, largely uncontrolled streams.

In many cases, truthful online information does not have much competition. Garbage collection *is* on Wednesday. Google Maps generally *does* get you from A to B. Your weather app *will* often correctly predict tomorrow's temperature. However, in cases where opinions vary, or where there is opportunity for sensationalism, misinformation abounds. Part of it is *dis*information, which entails messages that are *deliberately* misleading. Motives for doing this vary but are unlikely to be honourable. It could serve the purpose of scamming, for example, or be used for political suppression, or it could be a hostile act from another country (reporting of which might itself either be truthful or disinformation).

In part, misinformation spreads so easily simply because online messages in general come in enormous volumes. In 2022 and on average, people and robots shared 1.7 million bits of content on Facebook and sent 16 million text messages, nearly 350,000 tweets and 2.4 million snaps *per minute*.[16] However, misinformation also spreads *more easily* than truthful information. Soroush Vosoughi and his colleagues looked at Twitter and found that "falsehood diffuses significantly farther, faster, deeper, and more broadly than the truth in

Misinformation and disinformation

all categories of information, and in many cases by an order of magnitude".[17] They found that "many more people retweeted falsehood than they did the truth" and that "it took the truth about six times as long as falsehood to reach 1500 people".[18] In the field of GMO, Qian Xu and her colleagues looked at China's Weibo and found that there, too, falsehood spread faster and wider than truthful information.[19] Like Vosoughi, Xu also saw a people-factor, because the dissemination networks for falsehood were larger and more decentralised than the networks for truthful messages, which were smaller and consisted mainly of a few large sub-groups (which meant that the latter virtual network has less 'structural stability', in network theory speak). In addition, she concluded that falsehoods were frequently reposted by bots or cyborgs.[20]

Misinformation is a problem. Take Covid-19. In the 75 days from the beginning of January until mid-March 2020, when the virus had only just been discovered, over 240 million digital and social-media Covid-19 messages were sent around the world.[21] The spread and consumption of misconceptions around the virus were so instant and widespread that WHO already called it an 'infodemic' on February 2nd – less than two months after 'patient zero' had been confirmed in Wuhan.[22] In high-income countries and the world's urban centres, the virus was hotly debated. In these 75 days I spent time in the UK, the Netherlands and urban Kenya, where I found myself in frequent conversations about distancing measures. Some rigorous action was already being taken (I remember the first time it affected my own life in a significant manner, on 5 March, when a Dutch client decided to cancel all scheduled travel).

Rural regions in the Global South took much longer to notice this new virus. In these same 75 days I also spent two weeks in rural India and rural Ethiopia, and I do not recall a single conversation about the virus there. A mere month later these regions were suffering greatly. The virus itself was starting to cause death and disease in all parts of the world, but for the rural Global South its indirect effects were even worse. Health and other services had closed, and so had many markets. Group efforts such as fishing expeditions were curtailed and public gatherings had been forbidden – even for key events such as weddings and funerals. Border communities had lost their economic and nutritional lifelines as borders had been closed (see Box 2.1 in Chapter 2 for an illustration). In a 2022 review on peacebuilding I found that Covid-19 had compromised the peace process in Colombia as it had meant that "site visits from the government and international organisations were no longer possible, and this widened the space for armed groups to take control of some rural regions".[23]

When something invisible – such as a virus – causes despair, fake news spreads particularly rapidly, if very unevenly,[24] across the world. Even in today's global village, such streams are largely region-specific, and the nature of the falsehoods differ. In China, social media promoted a range of ineffective measures to prevent Covid-19 (Vitamin C, fireworks) or to cure it (oseltamivir, ritonavir).[25] In Iran, hundreds of people died because of acute methanol intoxication, commonly found in Iran's illegally produced alcohol, which people thought served as a prophylaxis.[26] In Kenya hard liquor was also thought to have a disinfecting effect, but the more prominent fake news was that hot weather prevented the spread of the virus.[27] The same hot-weather fiction appeared in Sudan, and if people did get Covid-19 then antibiotics was thought to cure them.[28] In Tanzania, people thought they could cure Covid-19 by inhaling hot steam (a message famously endorsed by the late president)[29] and according to Zambian social media people with black skin could not get the virus.[30] In India, Covid-19 led to the consumption of cow urine after clips of a 'Covid-curing' cow urine drinking festival went viral.[31] In rural Bangladesh, misinformation included the notion that Covid-19 could be spread by flies and mosquitoes and that Covid-infected people breathed out a skin-damaging gas.[32] In Nigeria some people believed that rural communities were not at risk as Covid-19 only affected people of higher socio-economic status.[33] And so forth.

Much of the misinformation led to a belief that the *actually* effective preventative measures were unnecessary. In addition, it sometimes aggravated the stigmatisation of marginalised groups or confirmed the hostility towards neighbouring countries. In India, lower-caste groups, Muslims and the rural poor population at large were widely thought to have high rates of infection and to be a threat to others (a belief that was factually untrue).[34] In Bangladesh, social media reported that India paid people to spread the disease in Bangladesh.[35] It also led to *overly* cautious behaviour. In West Africa in particular, there were streams of misinformation that warned against interaction with animals, which unnecessarily hampered the livelihoods of hunters and livestock farmers. Often, governments tried to counter misinformation, but in the latter case a number of West African governments and indeed WHO itself contributed to the spread of misinformation, by issuing public statements that included the following:

> "[Covid-19] has been found in many animals including bat, cat, camel and cattle" (Liberia); [...] "No unprotected contact with live wild or farm animals" (Nigeria); "Thoroughly cook meat, fish and eggs before eating them" (Mali); "Avoid unprotected contact with wild or breed animals and with surfaces in contact with them"

(Togo); "When visiting fairs and markets, avoid direct contact with live animals and surfaces where live animals have been handled" (Guinea Bissau). These messages were illustrated with drawings of animals and/or food (bats, chicken, pigs, ham, steaks, fish, eggs, pangolins, grasscutters, or snakes). [...] The WHO Regional Office for Africa [...] advise[d] to: "Avoid raw meat" [...] and to "Avoid direct contact with live animals. If this is impossible, make sure to clean your hands afterwards" (illustrated with a chicken and a pig). [...] These messages have no scientific basis.[36]

Once people enter a certain stream of online misinformation it is hard for them to get out of it, as algorithms spot their apparent interests and offer them more of the same. In rural contexts with limited access to digital messages there is another trap, which occurs when snippets of digital misinformation are spread orally. Those that resonate with local belief systems are assumed to be credible, and they may be seen as truthful for a long time as people do not have access to channels that convey corrective messages.[37]

Even at the height of the pandemic, misinformation caused millions of rural people not to believe that Covid-19 existed in their environment. It also caused them to take ineffective and sometimes harmful preventative and curative measures, and to stigmatise marginalised groups and other countries they saw as guilty of spreading the disease. Disastrous as its effects have been, misinformation around Covid-19 is merely an example. Other large streams of medical misinformation relate to blood transfusions and vaccinations against common illnesses. Marketing disinformation leads rural people down the path towards gambling disorders and unserviceable debts. Political disinformation, including from governments themselves, thwarts people's ability to make informed election choices. Social media companies remove some misinformation, but when Lisa Garbe and her colleagues compared Facebook's and Twitter's removal actions in a few countries they found large differences and concluded that these companies "hardly removed any content in Africa".[38]

Farmers are also deeply affected by misinformation that is specific to agriculture. Misinformation about pesticides was already common before the advent of the internet[39] and is still part and parcel of face-to-face advice as well[40] – and in recent years online misinformation has been worsening the problem. Farmers are unlikely to reap the benefits of drought-resistant or nutrient-rich GM crops because of GM-averse policies and legislation that are rooted in misinformation; and they may lose their markets where rumours spread that cow milk, chicken and tilapia could spread Covid-19,[41] or that grapes are injected with birth control pills,[42] or that aubergine causes cancer.[43] Online misinformation related to climate change, organic food and animal welfare also seems likely to indirectly affect rural life, though to my knowledge these linkages have not yet been researched. More generally it strikes me that, as of 2023, there is virtually no research on the impact of online misinformation, either mass-consumed or provided by agricultural extension centres and trade apps, on agricultural practices.[44]

Back in 2013, the World Economic Forum identified "the rapid spread of misinformation online" as one of the top ten trends facing the world.[45] Since then, the volume of online and social media misinformation has continued its exponential expansion. Recent global disruptions such as Covid-19 and the Ukraine war have generated enormous new streams of online misinformation and disinformation.[46] Governments around the world use the threat of misinformation as a justification for clamping down on people's liberty to voice their opinions. Such measures have not reversed the avalanche of misinformation, but they *have* shrunk personal freedoms and 'civic space' – which is the focus of the next and final chapter.

Notes

1   For an overview of advocacy voices in this field, their victories, the arguments they use, and the analysis as to why these arguments are weak, see Paarlberg, R. (2014) "A dubious success: the NGO campaign against GMOs", *GM Crops & Foods*, volume 5, issue 3, pages 223–228. For a balanced consideration of evidence, see Freedman, D.H. (September 2013) "Are engineered foods evil?", *Scientific American*, volume 309, issue 3, pages 80–85.

2   For a review of arguments used by both sides, but reaching a negative verdict, see Al-Attar, M. (2017) "Food ethics: a critique of some Islamic perspectives on genetically modified food", *Zygon*, volume 52, issue 1, pages 53–75; and Christian Ecology Link (undated) *GM crops? A Christian response*, Green Christian. For rare examples of cautiously positive papers, see Mahmod, S.S. and Kabbashi, N.A. (2013) "Ethical evaluation of GMO from an Islamic perspective", *3rd International Conference on Engineering Professional Ethics and Education 2013 (ICEPEE'13)*; and Idris, S.H., Abdul Majid, A.B. and Chang, L.W. (2020) "Beyond halal: maqasid al-shari'ah to assess bioethical issues arising from genetically modified crops", *Science and Engineering Ethics*, volume 26, pages 1463–1476.

3   Gomiero, T. (October 2018) "Agriculture and degrowth: state of the art and assessment of organic and biotech-based agriculture from a degrowth perspective", *Journal of Cleaner Production*, volume 197, pages 1823–1839.

4   For example, Oxfam UK (November 1999) *Genetically modified crops, world trade and food security; position paper*, Oxfam. There are also a few organisations that do *not* oppose GM crops but rather advocate for rules related to appropriate patenting, trade and consumer information, but in my experience these are rare. See, for example, World Vision Australia (April 2010) *Genetically modified food: an answer to food security?*, World Vision.

5 Both are covered in the chapter titled "Keeping genetically engineered crops out of Africa", in Paarlberg, R. (2009) *Starved for science*, Harvard University Press.

6 For many examples of human action leading to the development or prominence of certain weeds and pests (e.g., antibiotic-resistant bacteria, flies, body lice, caterpillars, red spider mites, fire ants, multiflora roses, mosquitoes and malarial parasites, the *Bt* bacteria), see Tenner, E. (1997) *Why things bite back: technology and the revenge of unintended consequences*, Vintage.

7 As a case study of how pests may "respond to the application of insecticides, and other management practices, by developing resistance", see Gassmann, A.J. *et al* (2012) "Western corn rootworm and Bt maize; challenges of pest resistance in the field", *GM Crops & Food*, volume 3, number 3, pages 235–244, with the quotation from page 236.

8 Qaim, M. (November 2010) "Benefits of genetically modified crops for the poor: household income, nutrition and health", *New Biotechnology*, volume 27, number 5, pages 552–557.

9 For a thorough assessment of GMO risks and uncertainties, and recommendations to fill gaps in safety assessments, increase regulatory clarity (rather than complexity) and improve innovations in and access to GMO technology, see National Academies of Sciences, Engineering, and Medicine (2016) *Genetically engineered crops: experiences and prospects*, National Academies Press.

10 See, for example, Shiva, V. (2006, fourth edition) *Seeds of suicide: the ecological and human costs of seed monopolies and globalisation of agriculture*, Navdanya. High-profile figures, including the UK's King Charles (before his accession to the throne), have repeated her claims.

11 Kloor, K. (Winter 2014) "The GMO-suicide myth", *Issues in Science and Technology*, volume 30, issue 2, pages 65–78.

12 The quotation is from page 1836 of Gomiero, T. (October 2018) "Agriculture and degrowth: state of the art and assessment of organic and biotech-based agriculture from a degrowth perspective", *Journal of Cleaner Production*, volume 197, pages 1823–1839.

13 I am not just talking about elite urban private schools (though I have seen it there as well): there are private schools in rural areas too, and sometimes they are the only available option. Also, many rural parents send their children to urban boarding schools, which sometimes use this pricing trap.

14 For example: "Monsanto did this in 1991 for a disease-resistant GM sweet potato in Kenya; Bayer AG, Monsanto Co., Orynova BV, and Zeneca Mogen BV did this for a high beta-carotene Golden Rice technology in 2000; and DuPont/Pioneer did this for nutritionally enhanced sorghum in Africa in 2005." Paarlberg, R. (2009) *Starved for science*, Harvard University Press, with the quotation from page 115.

15 *Ibid.*

16 Domo infographic 10.0, at https://www.domo.com/data-never-sleeps#, accessed on 7 September 2023.

17 Vosoughi, S., Roy, D. and Aral, S. (2018) "The spread of true and false news online", *Science*, volume 359, pages 1146–1151, with the quotation from the abstract on page 1146.

18 *Ibid*, with both quotations from page 1148.

19 Xu, Q. (2021) "Are you passing along something true or false? Dissemination of social media messages about genetically modified organisms", *Public Understanding of Science* (or 'PUS'), volume 30, issue 3, pages 285–301.

20 Bots are algorithm-controlled accounts that find and repost certain messages. Cyborgs do the same but are subject to some level of human control. Both work to maximise the visibility of certain types of messages.

21 Based on social media monitoring from The Vaccine Confidence Project, reported in Larson, H.J. (16 April 2020) "A lack of information can become misinformation", *Nature*, volume 580, page 306.

22 WHO first made this observation when the existence of the virus had only just been announced and WHO still called it 2019-nCoV: "The 2019-nCoV outbreak and response has been accompanied by a massive 'infodemic' – an overabundance of information – some accurate and some not – that makes it hard for people to find trustworthy sources and reliable guidance when they need it". World Health Organization (2 February 2020) *Novel Coronavirus (2019-nCoV); Situation Report 13*, 7 pages, with the quotation from page 2.

23 ICAI (December 2022) *The UK's approaches to peacebuilding*, International Commission for Aid Impact, with the quotation from paragraph 4.16 on page 14.

24 For the variables that affected the volume of fake Covid-19 news, which varied enormously across countries, see Shirish, A., Srivastava, S.C. and Chandra, S. (2021) "Impact of mobile connectivity and freedom on fake news propensity during the Covid-19 pandemic: a cross-country empirical examination", *European Journal of Information Systems*, volume 30, number 3, pages 322–341.

25 Chen, K. *et al* (2021) "Characteristics of misinformation spreading on social media during the Covid-19 outbreak in China: a descriptive analysis", *Risk Management and Healthcare Policy*, volume 14, pages 1869–1879.

26 Soltaninejad, K. (2020) "Methanol mass poisoning outbreak, a consequence of Covid-19 pandemic and misleading messages on social media", *The International Journal of Occupational and Environmental Medicine*, volume 11, issue 3, pages 148–150. Different sources give different estimates of casualties, ranging between 250 and 1,000.

27 Agade, H. (August 2020) *Misconceptions on Covid-19 persists in Kenya's rural areas*, CGTN Africa (for the hot weather argument) and BBC (April 2020) *Coronavirus: What misinformation has spread in Africa?*, BBC News (for the liquor argument).

28 Madyun, A. and Abdul, L. (June 2020) *Combatting myths and misinformation at Sudan's Covid-19 hotline call centre*, UNICEF Sudan.

29 BBC (April 2020) *Coronavirus: What misinformation has spread in Africa?*, BBC News.

30 Reuters (March 2020) *False claim: African skin resists the coronavirus*, Everythingnews.

31 The Week (2 April 2020) *Gujarat sees increased demand for cow urine amid Covid-19 scare*, The Week.

32 Sultana, S. and Fussell, S.R. (October 2021) "Dissemination, situated fact-checking, and social effects of misinformation among rural Bangladeshi villagers during the Covid-19 pandemic", *PACM on Human Computer Interaction*, Volume 5, CSCW2, Article 436, 30 pages.

33 Reported on page 455 of Okereke, M. *et al* (2021) "Covid-19 misinformation and infodemic in rural Africa", *American Journal of Tropical Medicine and Hygiene*, volume 104, issue 2, pages 453–456.

34 Islam, A. *et al* (2021) "Stigma and misconceptions in the time of the Covid-19 pandemic: a field experiment in India", *Social Science & Medicine*, volume 278, 9 pages.

35 Sultana, S. and Fussell, S.R. (October 2021) "Dissemination, situated fact-checking, and social effects of misinformation among rural Bangladeshi villagers during the Covid-19 pandemic", *PACM on Human Computer Interaction*, volume 5, CSCW2, Article 436, 30 pages.

36 Seytre, B. (2020) "Erroneous communication messages on Covid-19 in Africa", *American Journal of Tropical Medicine and Hygiene*, volume 103, issue 2, pages 587–589, with the quotation from page 587.

37 By means of illustration: such processes are reported in detail for three villages in Bangladesh, in Sultana, S. and Fussell, S.R. (October 2021) "Dissemination, situated fact-checking, and social effects of misinformation among rural Bangladeshi villagers during the Covid-19 pandemic", *PACM on Human Computer Interaction*, volume 5, CSCW2, Article 436, 30 pages.

38 Garb, L., Selvik, L.M. and Lemaire, P. (2023) "How African countries respond to fake news and hate speech", *Information, Communication & Society*, volume 26, issue 1, pages 86–103, with the quotation from page 88.

39 There is a long-standing phenomenon in which "farmers often face both information shortage and misinformation, with little awareness of environmental effects of pesticides. These factors may lead to pesticide overuse and misuse from framers and result in agroecological and environmental damage, which further creates 'lock-in' effect[s] of pesticide dependence at the farmer level". This quotation is from page 8 of Hu, Z. (2020) "What socio-economic and political factors lead to global pesticide dependence? A critical review from a social science perspective", *International Journal of Environmental Research and Public Health*, volume 17, issue 21, 22 pages.

40 For an example that covers Ecuador, Peru and Bolivia and that found that only 12% of almost 10,000 pieces of pesticide advice was accurate, see: Struelens, Q.F. *et al* (2022) "Pesticide misuse among small Andean farmers stems from pervasive misinformation by retailers", *PLOS Sustainability and Transformation*, volume 1, issue 6.

41 Sultana, S. and Fussell, S.R. (October 2021) "Dissemination, situated fact-checking, and social effects of misinformation among rural Bangladeshi villagers during the Covid-19 pandemic", *PACM on Human Computer Interaction*, volume 5, CSCW2, Article 436, 30 pages. Some fish farmers responded to the tilapia rumour by poisoning their ponds to kill off their fish stock.

42 This is one of the ten case studies in Guo, L. (2020) "China's 'fake news' problem: exploring the spread of online rumors in the government-controlled news media", *Digital Journalism*, volume 8, issue 8, pages 992–1010.

43 This example is also from Bangladesh and is reported in Chowdhury, A. *et al* (2023) "Systematic review of misinformation in social and online media for the development of an analytical framework for agri-food sector", *Sustainability*, volume 15, article 4753, 26 pages. Look for the word "brinjal", which is another word for aubergine (or eggplant).

44 *Ibid*, for an attempt to make an inventory of research that turned out not to exist.

45 WEF (2013) *Outlook on the global agenda 2014*, World Economic Forum, with the issue covered on pages 28a–29b.

46 Compare the Domo infographic 1.0 of 2013 with 10.0 of 2022 at https://www.domo.com/data-never-sleeps#.

14 Inclusivity and civic space

Using India and Egypt as case studies, Chapter 6 covered the potential benefits of e-government. However, any such benefits only reach people who have access to the e-government system – and some groups are deliberately excluded. To return to the examples of India and Egypt:

- India's national digital ID system, Aadhaar, was hailed as ensuring "the basic right to an acknowledged existence".[1] However, the government of India excluded almost two million people in Assam on the grounds that they were foreigners,[2] and Payal Arora says that the very system itself has its roots in "a 1992 government campaign to deport undocumented Bangladeshi immigrants through the tracking ability of the biometric identity database".[3] (Another ICT application, Facebook, served as an amplifier of hate speech that egged on discriminatory practices during the citizenship count in Assam.[4])
- In Egypt, Bahá'ís could not initially get digital IDs without lying about their religion, as the only three options were Islam, Christianity and Judaism.[5] This changed in April 2009, after a lengthy legal battle.[6]

As digital identification is often not merely a facilitator but a *requirement* to gain access to entitlements and services, or even just to participate in society, the fact that such identification may formally be voluntary is meaningless and the implications of exclusions are grave. Without digital identity, people may not be able to go to school, study, work in the formal sector, vote, buy goods from state shops or open bank accounts – to give but a few examples.

As far as I know, such deliberate group exclusions are rare. More commonly, governments use their ICT capabilities for surveillance purposes. Key parts of surveillance are tracking (following a person in real

DOI: 10.4324/9781003451716-17

time) and tracing (finding someone or retroactively checking one's movements and contacts). The pandemic illustrated the extent to which some governments use digital data for both. Vietnam applied its digital surveillance capabilities to prevent Covid-19 from spreading within its borders – and until mid-2021 it did so very successfully.[7] In the words of Bill Hayton and a Vietnamese author who writes under the pseudonym of Tro Ly Ngheo:

> When the Hanoi-based economic consultant Raymond Mallon returned home after a trip abroad in late March, he was immediately texted by the local police asking after his health. Vietnam is a state that not only knows where you live but also knows when you go away – and your mobile phone number. The degree of control matters because Vietnam has been widely praised for its success in tackling Covid-19. As of May 12 [2021], the country had, according to official statistics, suffered no deaths from the virus and had limited total infections to just 288, despite being next door to China and a popular holiday destination during the spring festival, when the coronavirus first hit the Chinese city of Wuhan. This has led many observers to suggest that the country's pandemic control strategy could be a model for others to copy, especially for developing countries.[8]

Compared to other country governments in Asia, the Vietnam government's virtual capabilities are not particularly advanced,[9] but none of the African countries will be able to copy Vietnam's movement and contact tracing model in the near future.[10] At the time Raymond Mallon received his text messages, and Vietnam subsequently sealed its borders altogether, I crossed a minor rural border from Ethiopia to Kenya, and back, without any type of identity check. This was just the latest of many times I had crossed African borders and checkpoints in the company of people who carried no form of identification with them, or who presented an identity card that had its photograph scratched off so as to serve the needs of multiple individuals. This was rarely cause for nervousness as it was nothing a small bribe could not resolve (and at the border crossing I just mentioned it merely required a wave from the driver to get the barrier lifted). The figures confirm the vastness of the gap. Some 96% of the Vietnamese population has a digital ID and most citizens have a digital trail (i.e., are active social media users). Conversely, in the two most populous African countries – Nigeria and Ethiopia – only 28% and 35%, respectively, have an ID, and most of the people who do have one are not active on social media.[11] Still, for several African countries it is probably just a matter of time before similar movement monitoring becomes possible, as they are achieving rapid progress with the digitalisation of their societies – albeit from a very low baseline compared to most of the rest of the world.[12]

Country governments that do have surveillance capabilities that are comparable to Vietnam's could potentially follow its (and, after an initial delay, China's) approach to control epidemics. They could also use those capabilities to squash dissent. The just-quoted paper on Vietnam's approach to the pandemic says that:

> [...] tools of Communist Party control [in Vietnam] [...] have now been repurposed in the service of health protection. The same systems, born from the same roots, made it possible for China to eventually control its outbreak, even after thousands of people died.[13]

The systems that Vietnam and China used to deal with Covid-19 tracking are part of the countries' larger systems that monitor people's movements and activities. These systems do not *have* to be ICT-supported. When I naïvely crossed the Pakistan–Chinese border on a bicycle, back in 1993, and cycled into a remote and sparsely populated part of Xinjiang (or the 'Xinjiang Uyghur Autonomous Region'), it took the Chinese authorities a mere two days to find and interrogate me. But ICT does make China's tracing and surveillance capabilities far quicker and more comprehensive still. When mobile and smartphones were not yet common, the government of China closely monitored internet cafés. Customers had to be registered and their online activities had to be stored for at least 60 days – and between June and September 2002 alone the government closed down 150,000 unlicensed internet cafés.[14] Then, the rapid spread of smartphones meant that the social media contributions, phone conversations and whereabouts of even the remotest of rural people in China have become knowable in real time. This is because of phone tapping and GPS tracking, in combination with a facial recognition app called Zhen Ni ('the real you', which is integrated in WeChat, an app that people in China frequently use).[15] This does not mean that people are not active on social media, or that they do not post critical messages. In fact, the government has not been able to halt the internet-fuelled decline of the trust farmers have in their local government.[16] However, it does mean that the virtual world is not a safe space for them.

The collection and use of surveillance data are not subject to informed consent, and may be used to restrict a country's 'civic space' –

Sure, play chess...

the set of rules and practices that jointly shape the extent to which people are able to organise, participate and communicate without hindrance and, in doing so, are able to claim their rights and influence the political, economic and social structures around them.[17] Civicus – an NGO – categorises civic space in both China and Vietnam as 'closed', which is the worst of five possible civic space categories (i.e., open, narrowed, obstructed, repressed, closed). In a very light manner, the suppression of civic space manifested itself in my evaluative work: I found the limits regarding who I could speak with and what I could ask, and the nervousness around that, to be more stringent in China than in any of the other countries in which I have done evaluative work.[18] In a far more serious manner, this suppression manifests itself in the way these countries' governments, and many other governments, deal with dissenting voices – and since the turn of the century there has been an almost uninterrupted trend for the worse. In the ICAI review on DFID's partnerships with civil society organisations (CSOs, a term that includes both NGOs and less formal civil society groups), I wrote that

> [...] in many countries, CSOs are less free to operate now than they were a decade ago. Political rights and civil liberties are under threat in countries on all continents. [...] CSOs have faced an increase in administrative hurdles, arrests, confiscation of equipment, forced closure and violence.[19]

... but Big Brother is watching you

Other dissenting voices – from journalists, protestors, opposition politicians – face these same obstacles, as well as the risk of torture and death.[20]

Since that 2019 ICAI review, there has been some progress. DRC and Sudan, for example, both moved from 'closed' to 'repressed' civic space (though Sudan may return to 'closed', depending on the outcome of the 2023 armed conflict). However, there has been further decline in many other countries.[21] In part, a negative neighbourhood effect makes this a self-reinforcing process, because a country's government will find it is easier to give in to the temptation to restrict its civic space if its neighbours have restricted *their* civic space as well. One argument governments often use to clamp down on people's ability to voice their opinions is the need to fight misinformation. Indeed, misinformation may have disastrous consequences, and Chapter 13 used its effects on the spread of Covid-19 as an illustration. However, simply suppressing information because it might be incorrect is risky as well – as again illustrated by Covid-19:

> The epidemic began with a poignant example of potential life-saving information that was suppressed as a rumour. On 30 December, Li Wenliang, a young ophthalmologist in Wuhan posted a message to colleagues that tried to call attention to a severe acute respiratory

syndrome (SARS)-like illness that was brewing in his hospital. The Chinese government abruptly deleted the post, accusing Li of rumour-mongering. On 7 February, he died of Covid-19.[22]

Covid-19 itself has also been used as a justification for strengthening surveillance[23] and further restricting movements and liberties.[24] The highly visible failure of some of the liberal democracies' attempts to contain the virus added to the attraction of authoritarian political models and the perceived legitimacy of tight government control – though some of that lost its shine when China finally loosened its restrictions in December 2022 and Covid-19 spread rapidly.

Just as governments use ICT to suppress civic space, civil society actors use it to continue their activism and to stay safe while doing so. In the past, they may have crossed borders to safety and then transmitted their radio programmes back into the country. Today, they do not have to stay so close as online activism can be carried out from anywhere in the connected world. Within their home country, activists use WhatsApp and other apps to inform each other about, say, their safety or about movements of security forces. They expose human rights violations by uploading clips of atrocities in real time. They encrypt their communication. They use virtual private network (VPN) services to bypass censorship, hide their whereabouts and prevent AiTM attacks ('adversary-in-the-middle' attacks whereby, unknown to the other participants, a third party inserts itself into their communication to manipulate it). And they use Ushahidi ('testimony' or 'witness', in Swahili), an open-source crowdsourcing platform designed for people to share and map information of human rights and other violations in fragile contexts. It was developed in Kenya, to map election violence in 2007. It has since gained functionality and the app, and other apps like it, have been used in a range of contexts.

Some donor country governments have contributed to ICT innovations that recipient governments then abused for the purpose of restricting civic space (see Box 14.1). Other investments are meant to help strengthen activists' digital safety. For example, a group of NGOs recently received over €50 million from a European government, on the basis of a proposal related to the protection of civic space that quite prominently included the use of Ushahidi-type applications, as well as investments to bolster the digital security of these NGOs themselves.[25] Whilst I have seen quite a number of similar examples, these investments seem insufficient to counter the long trend of declining civic space. First, donor countries just don't have much influence on the civic space in other countries, not least because many governments in the latter group of countries have recently issued laws that restrict advocacy NGOs' access to international funding. Second, if donors do manage to continue funding operations, the results of these operations may be easy to kill off. The 'I paid a bribe' platform mentioned earlier, for example, was invented in India by the Janaagraha Centre for Citizenship and Democracy, which receives money from the Ford Foundation and the charitable wing of Citibank (among many other donors);[26] but when Chinese platforms copied the model the Chinese government closed them down soon after they had started.[27] And third, many of the European and north American donors express their commitments to unrestricted civic space in policy statements[28] but rarely make it one of their priorities. I saw a few donors that found ways to get their funding to NGOs in countries in which foreign grants were no longer allowed (by funding a branch across the border, for example, or by smuggling actual cash into the country), but these were minor charitable foundations rather than government donors. The latter group will not generally risk violating the spirit of host country legislation, as they typically value trade, military cooperation, regional stability and a host of other interests more than the state of another country's civil society. Following an assessment of DFID's and the wider UK government's approach, I concluded that:

> Our analysis of country-specific DFID responses [...] show one commonality: the approach is risk-averse. DFID would like to see civic space expand in Ethiopia and Bangladesh, but it is not the UK's first priority in either country. DFID will not venture into work that might risk the broader suite of UK interests in these countries. One inevitable consequence is a selective approach towards civic space influencing work: like most other donors, DFID Ethiopia will not risk public confrontations on politically sensitive issues such as LGBT+ rights, or the suppression of civil liberties in the Somali region of Ethiopia, and DFID Bangladesh's response to extrajudicial killings and disappearances in Bangladesh has been muted. Similarly, DFID does not stand in the way of state control by, for example, providing civil society actors with technical means that lower the risk of surveillance when communicating online using social networks. Instead, DFID and the FCO lobby against further restrictions quietly, and behind the scenes, and DFID uses non-confrontational programming approaches. This includes trust-building work between CSOs and public authorities, which assumes that government action to limit civic space is caused by a lack of trust, rather than a political choice.[29]

This donor's assumption that mere trust issues cause governments to restrict civic space and to monitor their citizens' actions and whereabouts is probably incorrect. However, there *is* some evidence that rural communities do trust governments with their digital identity – or at least they believe that the usefulness of a digital identity outweighs possible drawbacks. A survey in rural areas in Andhra Pradesh, Rajasthan and West Bengal found "87% approval for mandatory linking of *Aadhaar* to government services" and that "for private services, [the] corresponding figure is 77%".[30] Many of India's 30,000–50,000 enrolment points[31] were rural-based[32] and, more than their urban counterparts, India's rural people were happy to queue "in the forty-degree Celsius summer heat of north India" to get registered.[33] Their enthusiasm is understandable, as e-government is likely to save them a great deal of time, and Aadhaar gives them access to facilities such as local money points (through the previously mentioned micro-ATMs) that save many millions of people hours of travel and considerable expense in order to receive transfers. However, as also discussed above, these digital identities introduce major risks and dangers as well. There is a thin line between power *to* and power *over* people.[34] Much like mobile money inadvertently served as a Trojan horse for online gambling, rural people may not readily see the risks of their online presence and may continue to "take to Facebook with gusto, sharing their lives online in spite of intense state, corporate, and interpersonal surveillance"[35] – unless they reside in Assam, Xinjiang or any of the other largely rural regions where people are already ruthlessly repressed.[36]

Box 14.1 Donor investments in digital IDs may do harm

Institutional donors have made significant investments in digital capabilities that serve laudable aims. The EyePay iris-scanning technology that UNHCR uses to provide refugees with cash assistance, for example, has far lower operating costs and is less prone to fraud than the actual physical cash distributions I managed when I worked for the UNHCR in the 1990s.

However, these are risky investments. This particular use of iris-scanning technology is part of a broader UNHCR approach of identity digitalisation of the entire refugee population that is formally under its protection. This approach has long been supported by the (mostly Western) donor community, but raises ethical questions.[37] The most immediately obvious issue is that harm can be done because "on some occasions, UNHCR conducts biometric registrations together with host states, or in some cases simply supports the hosts to carry out registrations. Such practices raise concerns about 'function creep', which refers to the way in which data collected for one purpose (e.g., to address fraud in aid delivery) may end up being used for an entirely different purpose (e.g., surveillance to combat terrorism)".[38] If host states are not initially involved, they may gain control over such data later. In Afghanistan, the Taliban seized US-operated biometric devices, bolstered with fingerprints and iris scans, that included data of people who the Taliban is likely to see as collaborators.[39]

Notes

1 Quoted in Nilekani, N. and Shah, V. (2015) *Rebooting India; realizing a billion aspirations*, Penguin Books India, on page 7. (The quotation itself originates from before Aadhaar had started: Nilekani, N. (2008) *Imagining India: ideas for the new century*, Penguin Books India.)

2 Masiero, S. (12 September 2019) *A new layer of exclusion? Assam, Aadhaar and the NRC*, London School of Economics. (NRC stands for the National Register of Citizens.)

3 Arora, P. (2016) "The bottom of the data pyramid: big data and the Global South", *International Journal of Communication*, volume 10, pages 1681–1699, with the quotation from page 1684.

4 See Section VI on pages 19–22 of Berényi, K. (2020) "Mapping minorities' vulnerability to hate speech and denationalisation with a focus on East and Southeast Asia", *Statelessness & Citizenship Review*, volume 2, pages 5–23. Assam-centred hate speech on Facebook (where Bengali Muslims were called "parasites", "dogs", and "rapists" who needed to be "exterminated") was comparable to the hate speech targeted at Rohingya in Myanmar in previous years, suggesting that Facebook's efforts to deal with hate speech had been largely unsuccessful, in part because it struggled with non-English hate speech.

5 See US Department of State (2004) *International religious freedom report 2004*, US Department of State. In the pre-digital era, some Bahá'ís managed to get an ID card in which the line on "religion" was left empty, but the digital registration system did not initially allow for this.

6 In 2009, a Supreme Court ruled that it should be possible to put a dash as an entry for "religious affiliation", and the first Bahá'ís received their digital ID later in that same year. Respectively covered in BWNS (17 April 2009) "Egypt officially changes rules for ID cards", Bahá'í World News Service; and BWNS (14 August 2009) "First identification cards issued to Egyptian Baha'is using a 'dash' instead of religion", Bahá'í World News Service.

7 Vietnam had a mere 26,000 confirmed cases by mid-2021; then the numbers went up rapidly, to 11.6 million confirmed cases in September 2023. See the 'Worldometer', worldometers.info/coronavirus/country/viet-nam/.

8 These are the opening lines of Hayton, B. and Ngheo, T.L. (12 May 2020) "Vietnam's coronavirus success is built on repression; the Communist Party's tools of control made for effective virus-fighting weapons", *Foreign*

Policy (there is no page number in the text's online version; the initial title at the time of publication was merely "Success is built on repression".)

9 For an assessment of Vietnam's 'whole-of-government' digital approach, and how it compares to ten other Asian-Pacific countries, see Okeleke, K. (2020) *Advancing digital societies in Asia-Pacific: a whole-of-government approach*, GSMA.

10 Sean McDonald describes a rare African digital contact tracing initiative, which was an entirely unsuccessful and ethically problematic attempt in response to the Ebola crisis and was based on call data records in Sierra Leone and Guinea. See chapter 3 of McDonald, S. (2016) *Ebola: a big data disaster*, the Centre for Internet & Society.

11 White, O. *et al* (April 2019) *Digital identification: a key to inclusive growth*, McKinsey Global Institute, Exhibit E1 on page 3 and Exhibit 4 on page 27.

12 For a 2016–2018–2020 comparison within Africa, and between Africa and other regions, see the infograph on page xxvi of DESA (2020) *United Nations e-government survey 2020; digital government in the decade of action for sustainable development*, United Nations Department of Economic and Social Affairs.

13 Hayton, B. and Ngheo, T.L. (12 May 2020) *Vietnam's coronavirus success is built on repression; the Communist Party's tools of control made for effective virus-fighting weapons*, Foreign Policy (there are no page numbers in the text's online version).

14 Zheng, H. (2013) "Regulating the internet: China's law and practice", *Beijing Law Review*, volume 4, number 1, pages 37–41, with these facts mentioned on page 39.

15 Pascu, L. (2 December 2019) *China introduces facial recognition for WeChat transfers, mandatory biometric scans for SIM cards*, Biometricupdate.com. Note that facial recognition is one of several identification methods (other options are fingerprints, iris scans, voice recognition and vein biometrics), but it is the one that is best suited for surveillance and people tracing, by linking the recognition database with cameras.

16 Wei, S. and Lu, Y. (2023) "How does internet use affect the farmers' trust in local government: Evidence from China", *International Journal of Environmental Research and Public Health*, volume 20, issue 4, 17 pages.

17 This text is paraphrased from the definition presented by Civicus, in its website section titled "Monitor: tracking civic space", accessed on 7 September 2023.

18 I did not have the same experience during evaluative work in Vietnam.

19 ICAI (April 2019) *DFID's partnerships with civil society organisations; a performance review*, Independent Commission for Aid Impact, with the quotation from paragraph 3.7 on page 11.

20 For a recent overview of issues, see the annual Civicus "State of civil society report" reports. In 2022, Civicus changed these reports' format – they are now shorter and easier to read. See Civicus (June 2022) *State of civil society report 2022*, Civicus.

21 For a global and country-by-country overview of recent developments in civic space, see the section titled "Civicus monitor: Tracking civic space", on the Civicus website.

22 Larson, H.J. (16 April 2020) "A lack of information can become misinformation", *Nature*, volume 580, page 306.

23 Newell, B. (2021) "Introduction: surveillance and the Covid-19 pandemic: views from around the world", *Surveillance & Society*, volume 19, number 1, pages 81–84; see also that issue's five subsequent papers, on pages 85–113.

24 Jefferson, A.M. *et al* (2021) "Amplified vulnerabilities and reconfigured relations: Covid-19, torture prevention and human rights in the Global South", *State Crime Journal*, volume 10, number 1, pages 147–169.

25 Specifically, the one-page country overviews include, respectively for five different countries, plans to: "provide physical and digital protection", "provide grants and digital [...] protection", "help them protect themselves physically and digitally", "pay particular attention to digital and physical security" and "strengthen each other's capacity in [...] digital and physical security".

26 See any of the hyperlinks at the bottom of the page of janaagraha.org/donors, which are summary data of foreign contributions.

27 Not everybody agrees it is this simple, and Yuen Yuen Ang writes that "the assumption that IPAB failed in China as a result of authoritarian intolerance and the suppression of corruption reports is partial and even inaccurate". For a detailed account of how the Chinese authorities came to close down the platform, see Ang, Y.Y. (October 2014) "Authoritarian restraints on online activism revisited; why 'I-Paid-A-Bribe' worked in India but failed in China", *Comparative Politics*, volume 47, number 1, pages 21–40, with the quotation from page 22.

28 Commonly made statements are like this: "We will address declines in the operating space for civil society that reduce civil society's ability to improve the lives of poor people and hold those in power to account. [We] will support organisations that protect those under threat and increase understanding of the extent, causes and consequences of closing civic and civil society space." This particular quotation is from DFID (November 2016) *Civil society partnership review*, Department for International Development (now merged with FCO, into FCDO), page 10.

29 ICAI (April 2019) *DFID's partnerships with civil society organisations; a performance review*, Independent Commission for Aid Impact, with the quotation from paragraph 4.67 on page 31.

30 Abraham, R. *et al* (May 2018) *State of Aadhaar report, 2017–18*, IDinsight, with the quotation from page 1 of the executive summary and details about the survey on page 3 of the main report.

31 Different sources present different figures. The estimation of 30,000 enrolment points, staffed by 100,000 certified people, comes from Nilekani, N. and Shah, V. (2015) *Rebooting India; realizing a billion aspirations*, Penguin Books India, with the estimate presented on page 35; the 50,000 estimate comes from White, O. *et al* (April 2019) *Digital identification: a key to inclusive growth*, McKinsey Global Institute, page 93. A possible explanation for the discrepancy could be that 20,000 additional points might have been opened between 2015 and 2019.

32 For a description of how this was done, see the first few chapters of Nilekani, N. and Shah, V. (2015) *Rebooting India; realizing a billion aspirations*, Penguin Books India.

33 *Ibid*, pages 8–9.

34 For a discussion of the unhelpful way in which literature appears to be entrenched into two camps – one promoting 'datafication' as a means of

increasing the inclusion, recognition and empowerment of affected populations, and one rejecting it as 'technosolutionism' and 'technocolonialism',
see Weitzberg, K. *et al* (January–June 2021) "Between surveillance and
recognition: rethinking digital identity in aid", *Big Data & Society*, 7 pages.

35 This quotation is from page 4 of Arora, P. (2019) *The next billion users:
digital life beyond the West,* Harvard University Press.

36 Not everybody agrees with this analysis and Keren Weitzberg and her colleagues say that the assumption that people "simply do not understand
enough about data systems to be critical of them is patronising". Weitzberg,
K. *et al* (January–June 2021) "Between surveillance and recognition:
rethinking digital identity in aid", *Big Data & Society*, 7 pages, with the
quotation on page 3.

37 For an overview of the UNHCR identity digitalisation approach and an
assessment of the ethical problems this poses, see Madianou, M. (2019)
"The biometric assemblage: surveillance, experimentation, profit, and the
measuring of refugee bodies", *Television & New Media*, volume 20, issue 6,
pages 581–599. UNHCR responded to some of the issues raised, here and
elsewhere, in UNHCR (15 June 2021) *News comment: Statement on refugee
registration and data collection in Bangladesh*, United Nations Refugee
Agency.

38 *Ibid*, page 588.

39 Reported in, e.g., Chandran, R. (20 August 2021) *Afghan panic over digital
footprints spurs call for data collection rethink*, Thomson Reuters
Foundation.

Conclusions

In rural regions in large and expanding parts of the Global South, an increasing range of ICT products and services is on offer at increasingly affordable prices. There are well-evidenced instances of such products and services enabling farmers to make more profitable crop choices, because connected farmers face lower risks when growing more perishable crops and because new, ICT-powered microfinance options enable them to move to slower-maturing crops that are ultimately more profitable. There are also examples of farmers increasing their volume of production, as ICT products improve their access to better seeds, tools and advice; and of them fetching more stable and sometimes higher prices for their produce, as buyers converge or pay more, lest they lose their suppliers.

Mobile phones and apps help rural people find off-farm income opportunities as well, as they make it easier for demand for and supply of labour, products and services to find each other. The advent of mobile money increases the frequency and volume of remittances, which is positive, as the net benefits of remittances for households and their communities are well-evidenced in Africa, Asia and the Americas alike. Rural social assistance programmes are cheaper, safer and easier to implement now that they have moved to virtual money options, and where e-government works well it reduces red tape and corruption. Rural people stand to benefit from this even more than urbanites, as they would otherwise have to travel for longer and might be more vulnerable to corruption. In some of these cases, product uptake and behavioural change happen at a speed not seen before in the broad field of socio-economic development.

In other areas, early evidence is also promising. Index-based micro-insurance is facing a range of legal and practical obstacles, but it does appear to reduce distress sales and other forms of negative coping behaviour, and it encourages farmers to be more risk-taking (in a good way) in their crop choices and farming practice. Lockout technology

DOI: 10.4324/9781003451716-18

appears to incentivise people to pay what they owe without causing them to face the threat of unmanageable indebtedness. During the pandemic, when markets rapidly changed and transport routes were closed, trading apps helped farmers find markets for their produce.

These benefits, well-evidenced or not, appeal to donors, and have a re-energising effect. They come at a time when the results of traditional microcredit have proved to be disappointing and after many M4P programmes imploded soon after donor support for them had ended – to name just two of the disappointments that have plagued donors in recent times. There is a sense of momentum, too, as the world is increasingly well-connected and 2G, 3G and even 4G networks are penetrating increasingly deep into rural regions. Moreover, these benefits promise financial *sustainability*, especially where the private sector is driving initiatives, as the development of ICT products and services is often based on convincing business cases. For individual initiatives, donors hope that ODA seed funding will propel products into commercial viability – and DFID's investment in M-Pesa shows just how catalytic such seed funding can be. For the ICT sector in its entirety, donors hope that such proofs of concept may generate a demonstration effect vis-à-vis the commercial financial sector and that donors will, in due course, be able to gradually phase out their involvement. This has not happened yet, but donors believe that this is just a matter of time, and their Theories of Change optimistically predict that this turning point will come a mere few years from now.

The appeal of the perceived and actual effects of ICT is such that digital agriculture and other types of ICT-powered innovations are now a standard part of the strategies and vision documents that donors develop to achieve their equitable socio-economic development aims. This strategic pro-ICT choice is operationalised through grants, concessional loans, junior venture capital investments, awards and a range of other mechanisms. Through these various channels, donors subsidise the development and roll-out of a continuous flow of new ICT products and services. They are produced and marketed by start-ups, large multinational companies and NGOs. They are used by farmers, traders, civil servants and by us, today's and tomorrow's international development professionals. We enthusiastically use and contribute to all sorts of development-focused open access resources and open-source software. Our research and work benefit from ICT and this bolsters our confidence in the power of ICT solutions. This confidence reinforces the ICT focus of the international development sector and of the institutional donors that employ us, directly and via their many implementing partners.

The broad and increasingly visible degrowth movement has not managed to temper donor enthusiasm. Donors ignored the movement's concerns, partly because of ICT optimism but also because the degrowth vision is one of a shrinking world economy and this is *so* far removed from donor paradigms that the movement seems irrelevant. This needs to change, as the appeal of ICT products and services as vehicles for socio-economic development in the rural Global South does not fully stand up to scrutiny, and some of the movement's concerns are based on persuasive evidence.

First, the positive effects are not as unambiguous and sizeable as they are claimed to be by donors and the companies and NGOs they support. Base-of-the-pyramid ICT solutions are often presented as great equalisers, but in most countries rural men use them far more than rural women. Without rigorous action this will be the case for the next rural generation as well, as girls are growing up with far less access to ICT than boys. Moreover, positive findings in relation to, say, the effects of mobile phones on price stabilisation (such as in Jensen's Kerala fish trade research) or the benefits of mobile money (such as in Suri and Jack's M-Pesa research) are contested by subsequent research – also in studies that cover the very same regions, issues and companies (such as Srinivasan and Burrell's Kerala fish trade research and the M-Pesa research of Bateman and his colleagues). Publications from pro-ICT donors such as USAID and the World Bank are biased towards the more favourable research conclusions. They celebrate market apps and the benefits of anti-corruption platforms, for example, and ignore independent research that shows that market apps such as RML do not necessarily serve the interests of small producers, and that reporting platforms such as 'I paid a bribe' do not in fact reduce bribery. To my knowledge, *none* of the flagship publications of multilateral organisations have acknowledged that, at country and wider regional levels, the uptake of ICT products and services has not (yet) translated into notable changes in agricultural production trends, or in trends in G-CSPI or other poverty indicators, or in a deceleration of climate change or the deterioration of the world's natural resources.

Second, ICT products and services have significant negative effects. Not all advice-giving ICT apps improve agricultural production, as bad advice is bad advice, even if it looks slick and is based on big data. Such data also consolidate the power of a few massive transnational companies, and reduce the agency of the farmers that engage with them. Whilst there are a few applications that utilise and increase the visibility of farmers' indigenous knowledge, the net effect of ICT products and services reinforces the donor community's blind spot in relation to indigenous practice. This is because ICT applications tend to focus on a narrow range of high-value global crops rather than indigenous ones, and because indigenous knowledge is at risk of gradually being replaced by app-based instructions and 'nudges'.

ICT applications may substitute labour or benefit wealthy farmers far more than poorer ones. Marketing apps such as Khula's 'one big virtual farm' and the many farm-to-kitchen apps, as well as online buying apps such as OneBridge and Amazon, may improve household incomes among farmers and widen product choice in rural regions. However, this comes at the price of less lively physical rural markets and their economic spin-off effects, which in turn leads to increasingly less self-contained local economies. This is the very opposite of what the degrowth movement rightly argues is needed to reduce the agricultural carbon and environmental footprint; and the Covid-19-triggered border closures showed the dangers of being overly dependent on cross-border trade networks. Mobile money has great benefits, but has also brought new types of gambling opportunities that reach deep into rural regions, and this is likely to cause indebtedness on a massive scale. E-government saves remote rural people time and money, but it also reinforces a digital society that excludes people, such as Muslims in Assam and Bahá'ís in Egypt, and enables repressive governments to use digital data to shrink civic space. Governments do this in part on the grounds that they need to halt the spread of misinformation, which is indeed a massive problem that mushroomed after online and social media took off (though suppressing civic space is probably not an effective response to it). Misinformation reinforces the stigmatisation of marginalised groups, thwarts people's ability to make reasoned election choices, led to Covid-19 denial and the rejection of anti-Covid measures, delays or halts vaccinations against common illnesses and implodes markets for crops that are rumoured to be cancerous, Covid-causing, or inseminated with birth control medication – to give but a few examples.

Some of these drawbacks could have been anticipated, but others much less so and many of the examples used in this book illustrate that Melvin Kranzberg's 'First Law of Technology' applies squarely to the specific case of digital information and communication technology as well: "interaction with the social ecology is such that technical developments frequently have environmental, social, and human consequences that go far beyond the immediate purposes of the technical devices and practices themselves, and the same technology can have quite different results when introduced into different contexts or under different circumstances".[1]

This book argues that ICT products and services have substantial benefits but do not always live up to their promise. For donors, NGOs and social enterprises the dilemma is, or should be, that "digital technology [is a] force [that] can perhaps be guided and steered for certain purposes, but not necessarily fully controlled or employed",[2] and that the effects of ICT investments are therefore hard to predict. In my experience, there is insufficient awareness of this dilemma, and little time or effort is spent thinking of possible externalities of rural ICT investments. I have looked for but not found *any* discussion, from *any* relevant party, about the possible implications mobile money might have on gambling habits, for example, when mobile money initially took off.

To maximise the usefulness of rural ICT investments, and to minimise the risk of them causing harm, donors, NGOs and social enterprises could usefully do seven things.

1   **Consider costs and alternative ways to use funding.** ICT investments are fashionable, and recent donor and NGO strategies and Theories of Change often include digital investments as a matter of course. However, even if ICT investments are useful, they may not be *more* useful than equivalent investments in rural infrastructure, the roll-out of social assistance programmes, BRAC-type poverty graduation programmes or the development of drought-resistant crop varieties – to mention just a few examples of rural investments that may well offer better value for money.

Where ICT investments do seem cost-effective:

2   **Invest in ICT products and services that, in as far as we know now, are likely to have positive effects on poverty and inequality, and are unlikely to have substantial negative effects.** Financed with grants or junior debt and other forms of concessional financing, NGOs and social enterprises could (and do) invest in off-grid electricity supply, lockout technology, base-of-the-pyramid inclusion, apps that foster and utilise indigenous knowledge and that are crop-agnostic, and things that reduce the digital divide. Heeks's 'ladder of inclusive innovation' is a useful tool to compare initiatives against. So are criteria such as language and literacy requirements, as things don't work if people don't understand them. Note, in this context, that women are more likely to be illiterate than men, and less likely to speak national languages, and that ICT applications that do not consider this may deepen gender inequalities. Simplicity is also key to bridging the digital divide, and so is 2G functionality – even in regions with 4G coverage, as many of the rural poor

do not yet knowingly *have*, much less *use*, access to internet. In my experience, ICT-focused social enterprises and commercial companies are more likely to succeed than NGOs, whose ICT investments are often one-off projects that fall flat once project funding runs out and the NGOs move on to other projects.

3   As a non-financial contribution, advocate for easier international mobile money transfers to bring down the costs of international remittances. Success is possible: in the recent past, a joint government-NGO advocacy effort persuaded FATF to work towards easier international transfers for NGOs.

4   When making investments, have modest and realistic expectations, and be conscious that ICT developments may gain momentum quickly and that unintended effects are common and hard to control or confine. ICT 'solutions' are not generally a silver bullet for anything, and there are fields in which their usefulness is hyped beyond reason. ICT investments might help keep activists safe, for example, but they are not going to be able to expand civic space or reverse the trend of its shrinkage; and it is not useful, in the foreseeable future, to develop ICT solutions for CBNRM systems because the digital divide is too large for such systems to use more than the most basic ICT products. Most importantly, it is not possible to control the applications or limit the geographic boundaries of new ICT products and services. The August 2021 digital data capture by the Taliban is a recent example of an ICT risk that was ignored until it materialised. Product developers and their financers need to err on the side of caution, and they need to invest heavily in the remaining three items on this list of suggestions.

5   Evaluate ICT products and services. Micro-insurance is growing rapidly, but evidence of its effectiveness, and of the importance of design choices, is scant – and research that disaggregates effects on the basis of gender is even more unusual. Micro-pensions and micro-leasing initiatives have, to my knowledge, never (literally!) been thoroughly evaluated, or at least such evaluations are not in the public domain and I have not seen them in the course of my evaluative work. Many other recent ICT innovations remain unassessed and, as a consequence, we know little about the extent to which products truly work and hardly anything about their externalities. PharmAccess's aim to achieve a positive spiral of health insurance and clinic investments is very appealing, for example, but I am not aware of quality external assessments of the PharmAccess model's effects on rural health.

6   **Do more horizon scanning.** Even outside of the degrowth community, many authors before me have concluded that the "digital
transformation in agriculture and rural areas comes with a range of
(ethical) concerns", and several have "argued for a responsible
research and innovation (RRI) approach to digital transformation
in agriculture".[3] Notwithstanding this often-raised concern, action
has been slow and many innovations are initiated without meaningful horizon scanning and scenario development. The questions
ICT investors and developers need to ask themselves go well
beyond an innovation's direct aims and target groups. Many relevant questions even go beyond the effects of an individual investment, as investments can be individually useful but collectively
harmful. Are 'one virtual farm' models going to eliminate the physical local marketplace and its economic spin-off effects? Is digital
agriculture in Africa on its way to generating a new Lewis-type
rural-to-urban migration wave, where rural–urban migration is not
the consequence of meaningful income-earning opportunities in the
city but of an absence of income-earning opportunities on the farm
or in the village? Are agritech solutions that offer Facebook-type
models (i.e., that offer free or low-cost products in return for data
extraction) ultimately going to be the small farmers' downfall
because these 'solutions' enable, for example, micro-insurance
companies to insure only those farmers who are not in fact at risk?
Will overconfidence in big data analytics erode farmers' knowledge,
or the checks and balances needed to mitigate against the risk of
bad advice? Will efficiency gains meant to intensify agriculture leave
more land to nature, as ecomodernists argue it should? Or might
they lead to a rebound effect, whereby the more efficient use of
agricultural land does not lead to more space for nature but to
conversion of more nature to agricultural land? These are a mere
few of many questions about likely effects of ICT innovations, and
to date ICT developers have made little attempt to explore these
questions as part of their decision-making process.

7   **Mitigate against negative effects.** ICT developers and their financers
should not fear failure but should make sure to recognise it. They
must be willing to modify or pause or even to swiftly terminate
projects if risks appear to be materialising, and to mitigate damage
that has been done or is likely to be done. Patient investments in the
preservation and utilisation of indigenous knowledge are needed to
give counterweight to agritech's strong focus on high-value crops.
In parallel, governments have a key role to play. They should
regulate against farmers' company-dependency. Without such

regulation, large companies may use farm-specific data to their own rather than to the farmers' advantage; or use Nespresso-type set-ups in which farmers' initial investments effectively tie them to specific products; or deliberately 'black-box' their software so that only they can provide maintenance services and resolve problems. Regulation against online gambling is important as well, as are systems of gambling-related awareness raising and an expansion of harm reduction facilities to rural regions. Agricultural intensification efforts should be combined with functional zoning systems, to mitigate against the risk of rebound effects. And anything related to digital identification must be done transparently, and not infringe on people's privacy for the purpose of controlling them. ICT comes with great potential benefits, but it also comes with dangers, and the fate of Afghan 'collaborators' and people with an online gambling disorder show that the damage it causes cannot always be undone.

Notes

1 Kranzberg, M. (July 1986) "Technology and history: 'Kranzberg's laws'", *Technology and Culture*, volume 27, number 3, pages 544–560, with the quotation from pages 545–546.
2 Arts, K., Van der Wal, R. and Adams, W.M. (2015) "Digital technology and the conservation of nature", *Ambio*, volume 44, supplement 4, pages S661–S673, with the quotation from page S670. Koen Arts and his colleagues wrote this about ICT in conservation in particular, but the statement has wider relevance.
3 Rijswijk, K. *et al* (2021) "Digital transformation of agriculture and rural areas: A socio-cyber-physical system framework to support responsibilisation", *Journal of Rural Studies*, volume 85, pages 79–90, with the quotations from page 80. This RRI approach has four main principles: anticipation, inclusion, responsiveness, and reflexivity.

Annex 1: Methodology

Introduction

Rural regions in the Global South are facing serious problems. These problems – such as persistent poverty, climate change and natural resource deterioration – are 'wicked', not in the sense of being 'evil' but in the sense of being bewilderingly complex and tenacious. Their borders are unclear and ever-shifting. Their stakeholders are many, and these stakeholders' perspectives as to what the problems are, and what the solutions look like, differ dramatically. The inter-relatedness of these problems is such that attempting to solve any part of any of them may well create new problems elsewhere.

Coyan Tromp, a science philosopher whose book *Wicked philosophy* inspired me to conduct the PhD that led to this book, defines wicked problems as follows:

> [These are] issues that have turned out to be both persistent and resistant to easy solutions. They are not merely complicated, meaning that they have many components with many specific functions; they are truly complex: they are multi-level phenomena involving a multiplicity of mutually interacting actors and factors, and their functions cannot be localised in any specific component. We cannot simply combine the pieces to find out how the whole works. Moreover, many of these issues are intrinsically connected to each other on different, higher levels.[1]

The methods required to build a useful understanding of wicked problems have to do justice to the complexities and messiness of such problems. Disaggregating the challenges and considering their component parts in parallel silos is not enough. In reality, however, this is often what happens. Much empirical rural development research is implicitly rooted in positivism (which assumes that reality can be observed objectively), and in the context of wicked problems this leads to a narrow focus on what is unambiguously logical and measurable. This narrow focus leads to narrow solutions that deal with *elements* of these complex, wicked problems, while potentially making the overall problem worse. The narrow focus of conventional research is a serious limitation and may lead (and has in practice contributed) to "greed, monstrous overgrowth, war, tyranny, and pollution".[2]

Books on ICT and rural development run the risk of falling in that narrow-focus-trap. Indeed, many UN annual flagship publications include special ICT-themed issues, and these issues rarely truly explore Kranzberg's First Law of Technology which, as a reminder, states that:

> Technology's interaction with the social ecology is such that technical developments frequently have environmental, social, and human consequences that go far beyond the immediate purposes of the technical devices and practices themselves, and the same technology can have quite different results when introduced into different contexts or under different circumstances.[3]

In this annex I outline how I dealt with all this complexity.

1   It describes the overall analytical approach I used to develop an 'intersubjective narrative'.
2   It describes the inter-related types of work that, in many loops and iterations, fed into this book.
3   It reflects on biases in the data I used, and on the way my identity and my engagement with data, literature and theory may have coloured my research and this book.

The analytical approach used to develop the book's intersubjective narrative

This book uses three sources: the evaluations I have conducted over the past decade, public databases, and other researchers' empirical research and literature reviews. Together, these sources provided me with the data and insights needed to come to an overall narrative about the complex role of ICT in rural socio-economic development in the Global South.

These narratives incorporate many long-established inductive truths, such as that a rural girl is less likely to have access to ICT products than an urban man; as well as my own inductive reasoning, such as my generalisations about donor behaviour and results. I also regularly use deductive reasoning. For example, the Financial Action Task Force (FATF) proved to be receptive to arguments related to fairness and proportionality, so I think a concerted advocacy effort could lead to FATF reconsidering its advice on international remittances. In addition, and much more challenging, I developed these narratives, in part, by using *abduction*.[4]

Abduction (or *retroduction*, an equivalent term) is "non-valid logical reasoning where inferences are based on concomitance, on co-occurrences or striking similarities in behavioural patterns between the situation at hand and other, comparable situations".[5] Within the field of research methodology, the term abduction was introduced in the 19th century, by Charles Sanders Peirce.[6] It is used both as a method for developing hypotheses and as a method of recognising patterns across comparable but messy and complex situations to come to a coherent and persuasive narrative or 'plot'.[7] I used abduction for the latter purpose. I do not have strong evidence that advanced ICT cannot be useful in community-based natural resource management (CBNRM), for example, but it failed wherever I saw it and I present reasons, from within CBNRM and outside of it, that lead me to believe this likelihood of failure is generalisable.

Abduction combines the efficient use of data sets that are insufficiently complete to allow for straightforward inductive reasoning, with one's tacit knowledge of a field of work (Peirce calls this 'intuition'), to come to plausible points of view, hypotheses or, in my case, an overall narrative on the role of ICT in rural development in the Global South. I came to this narrative by using data from the three sources mentioned above, and by conducting and combining the results of two distinct types of work.

The data that informed this book

To come to the narrative presented in this book, I revisited my evaluation reports and data sets, and I reviewed literature and public databases. In this section, I cover these types of data in turn, as if I considered them separately. In the actual process I reviewed my and other people's data and reports in a series of loops, iterations and revisitations.

Revisiting my evaluation reports and evidence logs, time and again

Since the mid-1990s, I have worked in 45 countries: 35 countries in the Global South and 10 ODA-providing countries. The experience gained throughout these years shaped my thinking, and thus this book, in a manner that is not always attributable to specific countries or events. The link between this book and my work is stronger for the period starting in 2010. In that year I started my work as an independent evaluator in the broad field of international development. Since then, I led evaluations (and occasionally other types of assessments, such as needs assessments that underpin programme proposals) that brought me to 32 countries: Bangladesh, Belgium, Bosnia and Herzegovina, Cambodia, Cameroon, China, Colombia, Egypt, Ethiopia, Ghana, India, Indonesia, Jordan, Kenya, Laos, Lebanon, Liberia, Malawi, Malaysia, Mali, Nepal, Netherlands, Nigeria, Pakistan, Papua New Guinea, Rwanda, Sweden, Switzerland, Uganda, UK, Vietnam and Zambia.

Different evaluations required different methods. The most common components were preparatory literature reviews, documentation reviews, site visits, surveys and various types of interviews and focus group discussions. All components had something in common: if something seemed relevant, it was logged in an 'evidence log'. These evidence logs are evaluation-specific coding trees that are initially designed to match an evaluation's Terms of Reference and that then grow organically in the course of an evaluation. *Everything* of relevance goes into that log, duly coded and sorted. This means that it is always possible to show exactly what evidence an evaluative statement is based on.

To come to this book, I revisited these data sets and the reports they fed into. None of these datasets were entirely within the scope of this book, but nearly all of them included relevant material. So I read the logs' headings and went through the data under headings that seemed potentially relevant. As I read through these logs and reports, I took two types of notes.

First, I made concise inventories of evidence related to themes that seemed pertinent to the book. They looked like this (using just one example):

Effects on indigenous knowledge:

- Con: only key crops (TartanSense, Babban Gona etc. etc. etc. – almost all of them?); replace direct observation (weather apps, moisture measuring); hinder direct observation (eyes on screen, earplugs, anything else?)
- Pro: active farmer engagement (Yuktix, maybe non-agri like FB?). If app is crop-agnostic – sounds good but I've not seen use relevant to indigenous crops and forest products. CBNRM ICT also sounds good but stumbles in practice (e.g. REDD+ -

skill / awareness / incentives issue? ICT opportunities not covered in forestry conference in Cameroon – blind spot donors?)[8]

Second, I kept notes about tentative inferences across themes, which
was part of my process of abduction. Some of these notes merely covered observations in one field that seemed relevant to another field. My
learning about harm reduction in relation to drug abuse seemed relevant
to the growing problem of rural online gambling, for example. Other
inferences were based on so many observations that I dared to general-
ise. For example, donors' casual attitude towards (and lack of awareness about) ICT-related risks was so consistent that I believe it applies
to all types of donor portfolios in which ICT plays a role.

I put my notes in a rough order of what the narratives could look
like, and scored the weight of my evidence in order to identify where I
needed to supplement my data with thematic reviews of literature.
Much later, when writing the actual text, I returned regularly to my
logs and evaluation reports and used specific search terms (e.g., addic-
tion, insurance, distress sales) to remind myself of evidence. I used
Microsoft Finder for this, to cover all logs and reports in a single search
request. I also returned to the originals whenever I used an actual
example, to get the exact quotation or to double-check the context. In
the book, these examples are used to illustrate or clarify points, not to
substantiate them, as the latter is not normally possible with a single
example. I did this only when I felt an example would strengthen the
text, and not merely to showcase my in-field observations. In the particular case of reviews for the Independent Commission for Aid Impact
(ICAI), I did not have access to my evidence logs, as ICAI requires
teams to surrender their logs at the end of a review. This meant that I
only included data from ICAI reviews if either of two conditions were
met. If the published document reported on a finding, then this finding
was clearly not confidential. In such cases, I generally used direct quotations, to avoid inadvertently breaching confidentiality by deviating
from what was shared publicly. Alternatively, if I remembered relevant
data that were not captured in the report, I only reported on them if
these data had reached the public domain through other channels.

Reviewing literature and databases

If my own evidence base was sufficiently strong to make my points
independently, I wrote the text on the basis of my own findings. This
was the case for the donors' uncritical enthusiasm for ICT solutions, for
example. In such cases, I did not feel much need to substantiate my

points with other people's research. I would still use references, but only to provide specificity (to add that, by 2018, India had provided digital IDs to 98% of their populations, for example); to substantiate minor points (to confirm that the pandemic had not stopped the growth in the number and value of mobile money transactions, for example); or to position my line of argument within the wider literature.

Where my evidence base was less strong, I was pushed to interrogate some of the points that I could potentially make more fully. Do apps indeed reduce price volatility? Is digital tracking and tracing important in governments' efforts to shrink civic space? Is micro-insurance ever profitable? In such cases, I looked for literature on the issue and used it to refine, or sometimes altogether change, the points I was making. Contrary to my original expectations, for example, ICT has not caused a kink in the longitudinal crop production index in any country in the Global South. My search for relevant research generally started with a key word search in Google Scholar. If there were systematic literature reviews or public databases that provided what I needed, I would use those. Otherwise, I read the first few publications that seemed relevant, sometimes using their references to snowball to other research, until I felt that the point I was making – intact or revised – was sufficiently well-substantiated.

Engagement with data: identity, choices and biases

At best, my evaluations and the other sources I used enabled me to develop a coherent but subjective version of a messy, complex reality, rather than an account of reality that is objectively true. Notwithstanding the inevitable subjectivity of the findings, the overall narrative may be robust enough to offer insights into the type of policies, programmes, investments and agreements that might help overcome these wicked problems. Whilst these insights will never be objectively true, the evaluative process (explained later) did enable me to upgrade them from deeply personal and subjective to 'intersubjective' – still subjective but with a higher level of relevance for decision-makers. This upgrade was achieved by developing and negotiating these insights in rounds of consultations with a wide range of stakeholders. When it comes to wicked problems, good research conclusions are rarely independently reached: they are the result of a joint effort, in which non-academics play a vital role. They are also never 'right' but, at best, 'better than alternative conclusions'. They are temporary, too, because everything evolves – and little evolves as fast as ICT and its role in the world – and because sometimes the world changes profoundly as a consequence of unexpected shocks (or what Nassim Nicholas Taleb refers to as 'Black Swans').[9] ICT powers such shocks, but so did the Covid-19 pandemic and the Ukraine crisis. Tomorrow we might see a sudden collapse of intercontinental trade or a disease that decimates rice harvests: humankind has proven itself to be poorly equipped to predict what the next shocks might be, and where they might come from. I therefore hope that the conclusions and policy suggestions of this book are relevant today, but realise that some of them may well be outdated tomorrow.

In this section, I first explore the influences that shaped my evaluations, and how these evaluations in turn shaped my thinking. Then I reflect on the way I dealt with literature and with databases, and on their inherent biases. Lastly, I discuss the way I engaged with theory, and the advantages and drawbacks of my approach.

The influences that shaped my evaluations, and the way these evaluations shaped me

Coyan Tromp describes the 'transdisciplinary researcher' whose work might provide the ingredients needed for the development of a coherent narrative about complex global phenomena and challenges.[10] The evaluations I have conducted over the last decade have turned me into such a transdisciplinary researcher. Specifically:

- These evaluations typically drew from a range of disciplines and knowledge domains and built on perspectives from a wide range of stakeholders.
- They have been intersubjective (or 'reciprocally adequate'), because their findings were always refined as the result of multiple rounds of discussions with my teams, academic peer reviewers, donor representatives, NGO staff and the participants of the programmes that had been part of the evaluations' sample.
- All my recent evaluations have been portfolio evaluations: they do not focus on individual projects and programmes but on diverse donor portfolios that typically include several billion euros' worth of grants, allocated to hundreds of projects. Such evaluations encourage 'complexity thinking' in the sense that they cover complex systems or networks with lots of fuzzy, messy interdependencies and their challenge is to find, within all this complexity, opportunities to achieve progress.
- Notwithstanding this complexity, such evaluations follow a 'coherence approach' (or 'coherence theory of the truth').[11] This means that these evaluations were analyses that aimed to be internally

coherent and consistent and that built a meaningful narrative with a line of sight to recommendations that could help achieve the strategic aims these evaluations were assessing against.

All these evaluations are rooted in data sets. These data sets take the shape of elaborate 'evidence logs' that capture and categorise everything that I and the teams I work with see, read and hear about, in the course of an evaluation. These logs have limitations, as they are shaped in part by our identities, professional roles and some of the biases we have and cause in others. As I was always the one who set the categories and produced and categorised most of the evidence logs' contents, and since my team members were too diverse to allow for broad-brush descriptions,[12] the following bullet points focus only on my own identity.

- I am a well-educated middle-aged white man from the Netherlands and, as such, I have faced no discrimination and sometimes received preferential treatment. An example of this is my entry into the field of international development and the UN system, as a Junior Professional Officer (JPO), in the 1990s. This is a type of 'fast stream' position that comes with a sizeable professional development budget that is meant to accelerate one's learning, and such positions are almost exclusively filled by nationals from the handful of countries that can afford to finance them – such as the government of the Netherlands. My privileged access to this and other opportunities, and the relative absence of obstacles, probably colours my view of the world. It may have biased me towards optimism, for example, and towards an exaggerated belief in people's liberty to make choices – even in a world with a steadily declining civic space.
- This optimism bias might be reinforced by my evaluative experience, in two ways. First, evaluators are generally exposed to new, dynamic and exciting initiatives, simply because organisations do not contract evaluators to assess static situations. This means that my perception of the role of ICT applications in the rural Global South might be an overestimation of their *actual* role. Second, evaluative research is different from most other research because, in the words of Mosse (a researcher, evaluator *and* development practitioner): "Your thinking focuses not just on 'How do I understand this problem?' but also on 'How should we proceed?'".[13] This applies to 'formative' evaluators like myself in particular,[14] whose role is to find positive and useful ways forwards.

- My identity as a white man from the Netherlands also colours the ways people see me, and during evaluations people often incorrectly believed that I represented 'the donor'. On the one hand, my privileged position gave me easy access to development professionals, government officials, donors and NGOs, as well as to the women and men these stakeholders ultimately say they aim to support and empower (though there were exceptions, see Box 16.1). On the other hand, my (perceived) identity and influence often generated caution and a social desirability bias ("I should say 'yes' because he probably wants to hear that this project was successful"). Responses were also shaped by the awareness of many of the research participants that they had agency. They realised that their responses would co-shape the evaluation's conclusions and recommendations, and that this might either further or jeopardise their interests – and they strategised their responses accordingly.[15]

Box 16.1 The project elite in central Mali

When conducting evaluations in particularly fragile contexts, danger requires careful planning and compromise. This sometimes enables gatekeepers to keep me from meeting with relevant people. On such occasions, my research participants meet me in a safe location such as an approved hotel, which means that the interviews do not take place where communities are actually living and I cannot walk around and engage with people more spontaneously. In such cases, staging and selective access to community members are common.

I recently reviewed a few community-based projects in Mali. Much of the work was done in the rural central and northern parts of the country. I couldn't go there, said the donor's security officer, as these regions were unsafe and white foreigners like me were specifically targeted.

So we negotiated, and the compromise was that I would meet a group of 'representative members of the community' in a safe riverside hotel in the centre of Mali, to which I travelled in a convoy of three armoured vehicles. Upon arrival, the security officer checked the hotel while I remained in my bulletproof safe zone in the parking lot next to the hotel, from where I saw a foreign woman getting out of the river. Dressed in a towel, she casually walked along the hotel walls and disappeared through its gate. Maybe it wasn't *that* unsafe around here.

After a lengthy security inspection, I was let into a meeting room where I met with six people, with whom I talked for a few hours.

They were helpful and knowledgeable about these projects, and indeed about every other project implemented in that region.

In the course of the conversation, I came to understand that the reason they were so knowledgeable was that this little group of 'representative community members' consisted of a locally prominent woman, her son, her nephew, his wife, her sister, and the mayor of one of the larger villages. They all played key roles in the various aid projects in the region.

There are two things to learn. First, 'community participation' is at risk of 'elite capture': the monopolisation of aid flows and decision-making by a small group of powerful people. Second, my armoured convoy and the bathing foreigner illustrate that donors are very risk averse when it comes to the safety of their staff and associates. An implication of this overly cautious behaviour is that donors – and evaluators – often don't really know what's going on.

- Rural people in the Global South are culturally diverse, but none of these cultures are close to mine. My communication with them has generally been through an interpreter (and sometimes through two interpreters, if people spoke minority languages). In addition, most contacts were brief – generally between a few minutes and a few hours. Jointly, these limitations meant that, for me, it has been impossible to come to the *emic* approach followed by many anthropologists in particular (i.e., an 'insider' approach that can only develop within a community, or be adopted by outside researchers who truly immerse themselves in a community's life).[16] The implication of my inevitably *etic* approach is that I will often not have truly and fully understood the people I engaged with in the course of my evaluative work. For the same reason, an interpretivist approach (where I try to understand an individual's action and their environment by attempting to see it through their eyes) would not result in a great deal of personalised depth.
- Missing out on the intricate detail of people's mindsets meant that I only ever understood the larger points. This had obvious drawbacks. However, it also had an advantage, as it increased my ability, in each evaluation, to recognise patterns across comparable situations (the core of the *etic* approach and of abduction)[17] – even in the midst of complexity and in a wide range of contexts. Over time and across evaluations, the sum total of the patterns that have emerged from the many conversations I have had and observations I have made, have

helped me to map contemporary dynamics, challenges and opportunities in sizeable parts of the Global South. This enabled me to craft a map of a system ('the role of ICT in rural life') that co-exists, overlaps and interrelates with other systems (such as agricultural production and common resource management, for example).

- As an evaluator, I am predisposed to bottom-up transition thinking. This means that I attempt to identify practices and policies that have furthered development in one country or region, and that could possibly be adapted, adopted and scaled to generate, over time, changes in wider systems. For this contextualised replication and scale-up to stand a chance, the ideas behind these practices and policies need to be internalised by relevant stakeholders. This requires substantial engagement with donors, host governments and NGOs – also after an evaluation's analysis has been conducted. I spent a lot of time on such engagement. This went well beyond 'convincing' the various stakeholders: it is also a matter of discussing, refining and negotiating recommendations with, as the end result, a proposed way forward over which relevant stakeholders can feel a sense of ownership. In science philosophy jargon, my research takes a *mode 2 approach* that leads to intersubjective, negotiated insights, that were often in part co-created by my team members and evaluative counterparts.[18]

Engagement with literature

To come to the narratives told in this book, I freely used other people's published research, and I found the body of potentially relevant publications to be very diverse. Economists, agronomists, nutritionists, political scientists, anthropologists, demographers and logisticians all come to very different findings when researching agricultural production, for example, and all of their findings are potentially relevant. I am an economist by training, and economists will therefore attract my attention more instantly than, say, nutritionists. However, I am also an eclectic reader and used findings from publications that reported on empirical research from any field that struck me as relevant and that, in my subjective view, used sensible reasoning to interpret robust and generally recent evidence.

In all cases, I acknowledge that *anybody's* research, like my own, is coloured by assumptions and world views, by the research funder, and often by the paradigmatic bubble people are in. This applies to publications of the World Bank and United Nations as much as to, say, the *Journal of Cleaner Production* and *Third World Quarterly*. It applies to both a publication's analysis and the selection of its references. Sometimes, it is blatant.

> Zuckerberg laid out the economic benefits of helping the world's poor to join the digital age: "There was this Deloitte study that came out the other day," he told his audience, "that said if you could connect everyone in emerging markets, you could create more than 100 million jobs and bring a lot of people out of poverty." The Deloitte study, which did indeed say this, was commissioned by Facebook, based on data provided by Facebook, and was about Facebook.[19]

Generally, the bias is more subtle than this, and it often takes an effort to recognise it. Wherever I noticed a particular bias and felt it was relevant to the analysis, I referred to the origins of research publications, in the text or in an endnote, and to their underlying assumptions.

My referencing is densest in fields where the data from my own evaluations were relatively light. As mentioned before, I found many of my references via Google Scholar, typically starting with key words (e.g., "ICT indigenous knowledge") and then refining my search through Google Scholar's advanced search functionality (e.g., range of years of publication, searches within publications that referred to publication X, key words in title). I also did a lot of snowballing (i.e., reading X led to reading Y and Z), often with the aim to reach the publications that were rooted in primary research rather than the publications that merely repeated what others had concluded. As I did this, I never consciously used publications because they added credibility to my analysis while ignoring others because they did not.

Engagement with theory

This book is rooted in evaluations, observations and empirical research, which I combined into narratives that explain and deal with some key and evolving challenges and opportunities in the rural Global South. These narratives are not positioned within a single theory or paradigm and instead they present, in the words of Norman Long, "a form of theoretical agnosticism which some scholars would regard as verging on empiricism".[20] One implication is that this book presents lots of *ifs* and *buts* and often concludes that "sometimes it works and sometimes it doesn't". This makes the narratives a little unsmooth and uncertain, at times. However, the lack of stringent theoretical framing has a major advantage as well: it saves me from the trap of being predisposed to dismiss things as 'bad' or to praise them as 'good', depending on the paradigm I am following. Such predispositions are common in general and in the field of ICT in particular, and they lead to conclusions that tend to be predictable. Authors within the degrowth movement, or those who see a world dominated by *Empire*, are predisposed to see danger in agriculture-enhancing ICT applications, and likely to condemn whatever next major innovation originates from a large transnational company. On the other side of the spectrum, ecomodernists are predisposed to approve of the next agricultural ICT application that comes along, and are particularly enthusiastic about the next ICT initiative that incorporates small-scale farmers into long agricultural value chains. They are unlikely to give due weight to the risk of adverse externalities, or to the plight of those who lose, in relative or absolute terms, as a consequence of ICT innovations.

Neither approach works for an evaluator. We are theory-agnostic, to the extent possible. We look at interventions in relative isolation, assess their merits and drawbacks, and look for ways to enhance their effects. We aim to develop and present insights that are of use to institutional donors (and sometimes other stakeholders), not to theorists. This comes with the risk of not seeing the bigger picture, but it avoids prejudice as to what such a bigger picture looks like and, by implication, what our judgement of the next intervention is likely to be.

Notes

1 Tromp, C. (2017) *Wicked philosophy; philosophy of science and vision development for complex problems*, Amsterdam University Press, with the quotation from page 17.

2 Bateson, G. (1979) *Mind and nature: a necessary unity*, Ballantine, quoted in Lyon, T.J. (Summer 1980) "Mind and nature: a necessary unity by Gregory Bateson (review)", *Western American Literature*, volume 15, number 2, page 150 (with the quotation, obviously, from page 150).

3 Kranzberg, M. (July 1986) "Technology and history: 'Kranzberg's laws'", *Technology and Culture*, volume 27, number 3, pages 544–560, with the quotation from pages 545–546.

4 The word *abduction* is "based on Latin *ducere*, meaning 'to lead.' [...] The prefix *ab-* means 'away,' [so] you take *away* the best explanation in abduction." Merriam-Webster (undated) *Deduction' vs. 'Induction' vs. 'Abduction*, Merriam-Webster Usage Notes.

5 This definition is from page 188 of Tromp, C. (2017) *Wicked philosophy; philosophy of science and vision development for complex problems*, Amsterdam University Press. Tromp's definitions of abduction and retroduction are identical.

6 Santaella, L. (1997) *The development of Peirce's three types of reasoning: abduction, deduction and induction*, E-Paper of São Paulo Catholic University.

7 Worth, S.E. (2008) "Storytelling and narrative knowing: an examination of the epistemic benefits of well-told stories", *The Journal of Aesthetic Education*, volume 42, number 3, pages 42–56.

8 I use shorthand when taking notes, so this text reflects what I meant, not what I wrote.

9 For an outline of the dramatic influences that unexpected freak occurrences – 'Black Swans' – can have on the world, and for many examples of how a reliance on the analysis of past trends and events may not prepare us for the future, see Taleb, N.N. (2008) *The Black Swan; The impact of the highly improbable*, Penguin. Taleb's most pointed illustration is a graph that depicts a turkey's confidence in humankind, in the course of a life of daily feeds, until the day before Thanksgiving (page 41 of the kindle edition):

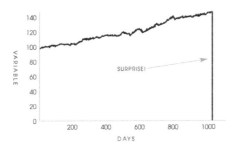

10 Tromp, C. (2017) *Wicked philosophy; philosophy of science and vision development for complex problems*, Amsterdam University Press.

11 Defined as: "The idea that the truth of knowledge depends on the degree to which statements about reality form a collective, consistent, coherent whole and provide a reliable explanatory theory." *Ibid*, with the quotation from page 189.

12 The members of my teams comprised 35 women and men in roughly equal numbers, with diverse backgrounds but mostly European, with ages that ranged between 23 and 65. Almost all team members had at least an MA-level degree, which they had almost invariably obtained from a European university.

13 Mosse, D. and Kruckenberg, L.J. (2017) "Beyond the ivory tower: researching development practice", Chapter 17 of Crawford, G. *et al, Understanding global development research: fieldwork issues, experiences and reflections*, Sage Publishing, with the quotation from page 4 of that chapter.

14 A 'summative' evaluator issues a verdict on a programme's value for money. A 'formative' evaluator makes recommendations that are meant to enable donors or implementers to improve their programme's or portfolio's value for money. My evaluative work is always formative.

15 The role of the (perceived) power of researchers, and of the importance of the nature of the identity of researchers and the way they relate to their research participants, is discussed in Crawford, G. *et al* (2017)

Understanding global development research: fieldwork issues, experiences and reflections, Sage Publishing.

16 The terms 'emic' and 'etic' come from linguistics, and specifically from Pike, K. (1967) *Language in relation to a unified theory of the structure of human behaviour*, Mouton. The *concept* of 'emic' already existed prior to Pike's 1967 publication and comes from anthropologists and ethnologists striving to understand culture from the native's point of view – popularised by Malinowski, B. (1922) *Argonauts of the Western Pacific*, Routledge and Kegan Paul.

17 "The emic is commonly concerned with particular ways of meaning-making and the etic with generalized comparable patterns." This quotation is from page 2 of Whitaker, E.M. (2017) "Emic and etic analysis", in Turner, B.S., editor, *The Wiley Blackwell encyclopedia of social theory*, John Wiley & Sons.

18 Rather than a *mode 1* approach of positivist analysis. For an overview of the initial concept of 'mode 2' knowledge and the way this concept evolved after its first years, see Nowotny, H., Scott, P. and Gibbons, M. (2003) "'Mode 2' revisited: the new production of knowledge", *Minerva*, Volume 41, number 3, pages 174–194. This was the introduction of the *Minerva* special issue on "Reflections on the new production of knowledge".

19 Bhatia, R. (12 May 2016) "The inside story of Facebook's biggest setback", *The Guardian long read*, The Guardian.

20 Long, N. (2001) *Development sociology; Actor perspectives*, Routledge, with the quotation from page 10. Long sees this as postmodernism.

Annex 2: Research integrity and ethics

Research ethics are the set of ethics that govern how research should be conducted, and how its findings should be disseminated. Research ethics are captured in codes of conduct. There are country codes, thematic codes for researchers in fields ranging from chemistry to psychology, and codes that are specific to institutions.

Research ethics principles apply to evaluative research as well. In addition, evaluative research is subject to two further principles. The first, for summative evaluations, is that they must hold stakeholders such as donors, NGOs or governments to account. The second, for formative evaluations, is that they should consider the possibility of recommendations that could be used to improve a project, programme, policy, portfolio, or whatever the focus of the evaluation might be.

I am from the Netherlands and in this annex I match my evaluative research practice against the principles outlined in the "Netherlands code of conduct for research integrity" (hereafter, 'the Code').[1] The first five sections of this annex cover the five principles of the latest (September 2018) version of the Code – honesty, scrupulousness, transparency, independence, and responsibility – beginning in each case with the definition of the term, quoted from the Code.[2] The section thereafter is based on the interviewing principles outlined in another document, the May 2018 version of the Dutch "Code of ethics for research in the social and behavioural sciences involving human participants" (hereafter 'the code of ethics').[3] In that section, I explain the ways in which I tried to follow these principles, and when and why I sometimes failed to do so. The final section reflects on the principle of maximising data utilisation.

As this book re-uses data that were gathered as part of evaluations, this annex discusses the operationalisation of the various ethical research principles as they applied *in these evaluations* rather than distinctly *for this book*.

194

Honesty

The Code's definition

> Honesty means, among other things, reporting the research process accurately, taking alternative opinions and counterarguments seriously, being open about margins of uncertainty, refraining from making unfounded claims, refraining from fabricating or falsifying data or sources and refraining from presenting results more favourably or unfavourably than they actually are.

Operationalisation

- *Accurate reporting on the research process.* Because people do not generally like being evaluated and because I need to maintain a constructive relationship in order to do my job, I spend significant time on relationship management. Most of this is a matter of keeping people in the loop on progress. This includes weekly or biweekly process updates, debriefings and 'back to office' notes at the end of country visits, and, towards the end of the evaluation process, one or more 'tentative findings' meetings. These meetings are the evaluative equivalent of the academic principle of communicative validation: I present and discuss findings before starting the report-drafting process.
- *Openness about margins of uncertainty.* I did not have a weighing system for evaluative findings until 2016, at which time I adopted an A-D scoring system.[4] Since that time, I score every finding, as follows:

  > *A* signifies a consistent finding across multiple sources, amounting to strong evidence of a systemic issue.
  > *B* signifies a finding that recurred frequently enough to suggest a likely systemic issue.
  > *C* signifies a finding that recurred on a number of occasions, but further research would be needed to determine its prevalence.
  > *D* signifies a single point finding, which is not sufficient to ground a broader conclusion.

I do not normally include findings that are scored as *C* or *D* in my evaluation reports, except in cases where a single data point was serious enough to ring alarm bells (such as in the case of sexual abuse or exploitation risks, fraud or significant misreporting). These ratings are shared, finding by finding, with external peer reviewers, and all the underpinning data are shared within the evaluation team and, in the case of reviews for the Independent Commission for Aid Impact (ICAI), with the ICAI Secretariat and relevant Commissioners.[5] In my 2020 evaluation of the €770 million Dutch Good Growth Fund, I also shared all preliminary findings with the Government of the Netherlands and the four implementing agencies – including the findings that did not make it into the report – with scores and references to indicate the weight and nature of evidence.

- *Consideration of alternative opinions and counterarguments.* Evidence is often ambiguous: it is common for there to be data points which confirm a finding, and data points which contradict it. In such cases, the A–D scoring of findings is based on the *weighing* of evidence. On the basis of this evidence, I consider alternative opinions and counterarguments in four stages.

  1 Once tentative findings are starting to emerge, the nature of interviews with all types of stakeholders changes and, where appropriate, includes a discussion of these emerging findings (at different conceptual levels, depending on the nature of the research participants).

  2 Then, in the course of 2–5 full days, the evaluation team collectively refines my initial findings. Findings that are insufficiently clearly embedded in data are modified or removed.

  3 Once the team has agreed on the evaluation's initial findings, external peer reviewers scrutinise these findings, and the evidence they are based on. Where the reviewers have a sufficient level of security clearance, they do this with access to the full evidence log.

  4 Once I have incorporated the peer reviewers' feedback, the findings are shared with the stakeholder that is being evaluated (typically an institutional donor, institutional investor, government or NGO). On the basis of an overview of findings (in which case every finding is coded A–D) or on the basis of a draft version of the report (in which case there is no coding of the evidence weight), this stakeholder makes factual corrections, offers counterarguments and provides new evidence.

- *Use of genuine data and sources only.* Where these are available, I use primary data and primary research reports (with the exception of literature reviews, which I use as starting points). I have never knowingly written incorrect notes; have never noticed team members doing this; and am not aware of an incentive that would entice us to do this.

Scrupulousness

The Code's definition

Scrupulousness means, among other things, using methods that are scientific or scholarly and exercising the best possible care in designing, undertaking, reporting and disseminating research.

Operationalisation

- *The scholarly nature of the methods and the care taken in design.* Most evaluations are guided by *ex-ante* peer-reviewed methodological frameworks. These frameworks serve three purposes:

  1 They allow for *ex-ante* methodological quality assurance.
  2 They minimise the imposition on relevant stakeholders by ensuring that the data-gathering process is designed with the research questions in mind (leading to focused and therefore efficient data gathering). This happens in two ways: the methods are closely linked to the research questions; and the framework provides a frame of reference that reduces our susceptibility to data-gathering drift (i.e., "oh, *that* sounds interesting!" – leading to tangential data gathering). In Annex 1 I explained that these methodological frameworks also contain some jeopardy, as they create the risk of a narrow evaluation focus. Such a narrow focus does not do justice to the complexities and messiness of problems, and may lead to narrow recommendations that deal with *elements* of these complex, wicked problems, while potentially making the overall problems worse.
  3 They help the organisations and people involved to prepare for the evaluation. This poses the risk of significant staging. We (the team) are generally able to mitigate this risk in two ways. First, we ensure that we choose the projects and sites that we visit on the basis of lists of options (i.e., we, not the implementers, make the selection). Second, we are generally able to reserve time for unscheduled additional visits and interviews and to improvise during our visits.

- *Care in undertaking research.* To maximise the quality of our research, we (my teams) follow fairly strict protocols (e.g., on how to start and end interviews) and follow principles such as those mentioned below (these are just four of many such steps).

  1 We conduct the first few interviews in teams of two, where possible, and reflect on the experience afterwards.
  2 We always meet with and brief interpreters before we start working.
  3 If we conduct our research in multiple locations and we are in regions with phone or internet coverage, we have daily evening-time virtual team catch-ups during country visits, to discuss findings and the next day's plans.
  4 Where we have reason to believe that a respondent has provided deliberately biased responses (such as a government official who uncritically toes the official government line, or a programme manager who presents an unrealistically positive account of the success of the programme), we discard notes from that interview. We also discard external evaluations if their methodological choices are unclear or indefensible, or if the analysis is not clearly sensible and evidence-based.

- *Care in research reporting* I type blind and fast, and most interview notes are close to verbatim. I nearly always finalise interview notes on the same day as the interview, which means there is minimum memory loss. I almost never record interviews, because the advantages of the recording (double-checking what was said, primary evidence to be presented in case of disagreements) do not outweigh the drawbacks (the risk of more reluctant respondents, additional preparation time spent at the start of the interview). My team members use other approaches to achieve the same note-taking rigour: they either record and transcribe interviews, or they conduct interviews in groups of two: one to conduct the interview, and one to take notes.

  All evidence, gathered from interviews, literature, documents and on-site observations, is logged in evidence logs that are based on evaluation-specific coding trees (see Annex 1). To the extent possible, the logging uses direct quotations from documents and verbatim interview notes, rather than interpretations.[6] Every database entry – from interview notes, documents or anywhere else – has a full reference.

  All evaluation reports are subject to external peer review, and to a fact-check from the subjects of the evaluations (which typically includes substantive feedback). Subsequent draft reports are

198

accompanied by my response to each of the comments (i.e., accept, partially accept or reject, with substantiation of the decision, including reference to relevant evidence, and an explanation of action taken).

- *Care in dissemination of findings.* Dissemination is never up to the evaluator: I merely send the reports to my clients – and, with their permission, to the research participants that had expressed an interest in the findings (final report versions only). The only exception is that I include, in my contracts, the stipulation that quoting from my reports requires my prior written approval.[7] I introduced this stipulation in 2013, after a client took a few positive statements from an otherwise very critical set of findings and used them for fundraising purposes.

Transparency

The Code's definition

Transparency means, among other things, ensuring that it is clear to others what data the research was based on, how the data were obtained, what and how results were achieved and what role was played by external stakeholders. If parts of the research or data are not to be made public, the researcher must provide a good account of why this is not possible. It must be evident, at least to peers, how the research was conducted and what the various phases of the research process were. At the very least, this means that the line of reasoning must be clear and that the steps in the research process must be verifiable.

Operationalisation

- *Clarity on nature, purpose and steps of the research process.* See the first bullet point in the section about scrupulousness, above.
- *Clarity of data sources, coding and sorting.* The following bullet points are not about this book but about its underpinning data sets.

  1 For people with appropriate security clearance, the evidence base of all findings of all evaluations is always clear, because *all* data are placed in a coding tree. If, for example, an evaluation covers the extent to which participating companies implement the Corporate Social Responsibility (CSR) standards of the OECD, I go to sub-topic "Q13a: Have the OECD CSR principles been applied by the participating companies?". If it is not there, we do not have it.

2   The data points in these (confidential) evidence logs are all fully referenced. For interview fragments this means the name and role of the respondents (and their sex in case a gender balance amongst respondents is relevant and a potential risk), the organisation they work for or programme they are part of, and the date and nature (e.g., face to face meeting, Skype or phone call) of the interview. For focus group discussion fragments this means the date and location of the discussion and an indication of the nature of the participants. For document quotations, it means the full reference, including page number and hyperlink, if a hyperlink exists.

- *Clarity in line of reasoning.* My 'tentative findings' template is a table with three columns. For each finding it presents, from left to right:

  1   The evaluative finding.
  2   The nature of the evidence (including counterevidence), presented in bullet point format in the sequence of the argument, with a database code reference to the data folder that contains the evidence that the statement is based on.
  3   The scoring of the weight (A–D), with, in case of a C or D, a brief indication of the nature of the evidence or evidence gap. (As mentioned above, C- and D-scoring findings are not normally incorporated in the report.)

- The rows are presented in the order of the build-up of the overall analysis (e.g., the findings are X and Y, which leads to conclusion Z). Table A2.1 provides an example (which is realistic but fictitious as the actual tentative findings documents are confidential).

- *Limits of transparency.*

  1   Most evaluation reports are publicly available for at least a few years.
  2   Raw data and 'tentative findings tables' are accessible to an evaluation's team members only, and sometimes to peer reviewers and staff members of scrutiny bodies – provided that they have appropriate security clearance. This means that, whilst I recognise the benefits of data transparency, also as a means to support fraud detection,[8] I am not in a position to comply with the FAIR principles[9] that the 25th standard of the Code refers to when it says that researchers should "contribute, where appropriate, towards making data findable, accessible, interoperable and reusable".[10]

Table A2.1 Fictitious example of a 'tentative finding'

| Project implementation is affected by donor-caused delays and uncertainty. | • P46 New projects almost invariably started with a delay of over six months compared to original timelines.
• P47 Projects that were meant to follow onto each other often faced 3- to 18-month gaps between phases, and this caused programme disruption and loss of gains.
• P47 In the few cases where there were no gaps between phases, grantees faced uncertainty until funding approval was granted only a few weeks prior to the next phase.
• P51 Most delays are caused by donor X, and by the multi-donor initiatives that donor X is part of.
• P51 Delays were poorly communicated with funding recipients.
• P51 Donor X sometimes attempts to reduce delays by imposing unrealistic timelines on recipient organisations. | A |

3   Because many of the data that underpin these evaluations are confidential, the evaluation reports and this book do not mention names and do not make statements that are traceable to specific individuals. In the case of ICAI reviews, I only used data that are presented in the publicly available ICAI reports or that are otherwise available in the public domain. In case of the latter, I truthfully present the finding as an ICAI finding but also provide the reference to explain why I am able to present the finding in this book.

Independence

The Code's definition

Independence means, among other things, not allowing the choice of method, the assessment of data, the weight attributed to alternative statements or the assessment of others' research or research proposals to be guided by non-scientific or non-scholarly considerations (e.g., those of a commercial or political nature). In this sense, independence also includes impartiality. Independence is required at all times in the design, conduct and reporting of research, although not necessarily in the choice of research topic and research question.

Operationalisation

- *Statement of principle.* Full independence does not exist in social research. Education, previous experience and operational environments all influence a researcher's mindset and choices (see Annex 1). The implication is that the bullet points in this sub-section are about 'independence to the best of my ability'.
- *Independence of sampling and methodological choices.* Evaluations generally start with an approach paper (sometimes named inception report or work plan). The client needs to check it before the work commences. Formally, the client's feedback is limited to fact-checking and should not affect the independence of my evaluations. In practice, this step has sometimes compromised independence. Specifically, I have often been asked to change my initial choice of country visits, generally for reasons of safety and security. The consequence is that my (and other evaluators') visits are often to the same group of relatively safe countries – such as Malawi, Jordan and Bangladesh, rather than Somalia, Syria and Afghanistan. I have no reason to assume that these arguments were anything other than genuine. I do not know the impact these changes may have made to my analyses, but note that evaluations tend to bias towards programmes in safer and more stable environments, where programme quality is likely to be higher and more closely monitored.
- *Independence of analysis.* The data analysis of the evaluations I lead are not 'independent' in the sense that I do not conduct the analyses by myself: see the sub-section on *Honesty*, above.
- *Independence of reporting.* Reporting is rarely entirely independent, for three reasons:

1 The subjects of evaluations have a vested interest in positive findings. Their feedback therefore tends to focus on countering critical findings.

2 I have sometimes been asked to keep confidential findings outside of the report. In such cases, I share these findings in a confidential post-assessment note. This note also often includes findings for which the evidence base is less than rigorous (Cs

and Ds): such findings are sometimes worth sharing, not as formal 'conclusions' but as 'points to consider'.

3   To mitigate the risk of misinformation and disinformation, reports are often checked for potential 'scandal value' and sentences that could be taken out of context and misused are modified.

Responsibility

The Code's definition

Responsibility means, among other things, acknowledging the fact that a researcher does not operate in isolation and hence taking into consideration – within reasonable limits – the legitimate interests of human and animal test subjects, as well as those of commissioning parties, funding bodies and the environment. Responsibility also means conducting research that is scientifically and/or societally relevant.

Operationalisation

- *A focus on research that is societally relevant.* I am self-employed and, because evaluation team leaders are in short supply, the demand for my services exceeds my availability. I am therefore in the fortunate position that I can afford to only take on work that has the potential to be useful. As a development economist, I consider an assignment to be 'useful' if it is likely to make a meaningful contribution to the socio-economic development results of ODA.
- *Consideration of the interests of 'human test subjects'.*

  1   **During the assessment process.** If the assessment protocol allows for this, I share my interview notes in cases where the issues discussed are particular sensitive, a day or so after the interview, and make corrections if corrections are sent to me. I never quote identifiable people without their explicit prior consent, and make sure that sensitive observations cannot be traced back to individual respondents. This is rarely difficult as evaluative observations that make it into reports are generally based on a number of sources and data points. (The next section, on interview principles, covers other elements of the interests of research participants.)

  2   **At the end of the assessment process.** At the very least, research should be aware of the risk of doing harm to research

participants, and should be explicit about its choices in this regard. In this context, my evaluative work is likely to cause many people harm, as evaluative findings often lead to programmatic changes, and these changes benefit some and harm others. For example, a recommendation to concentrate a fund's disbursements to five countries instead of the 20 countries may harm people in the 15 countries that are no longer eligible for the fund's disbursements. I consider this acceptable, provided that the net results are likely to be positive.

Interview principles

Design choices to reduce inconvenience, stress or embarrassment

When I interview people who are unfamiliar with formal interviews (or with speaking with foreigners) I try to minimise stress by following local conventions (e.g., sitting on the floor), using female interpreters whenever possible (especially when I think I will be interviewing women and girls), and using easy, non-confrontational methods, such as:

- The stimulated recall technique,[11] walking-while-talking[12] and other low-stress techniques that help form insights and show an interest in the respondent's experience without requiring the respondent to express opinions, pass judgements or maintain eye contact. This helps when, for example, assessing contentious issues that may be rooted in power differentials (e.g., why so many of the women who had applied to get a plot of land in an agricultural cooperative had been turned down; or why a distribution programme may have reached some groups more effectively than others).
- The nominal group technique,[13] which is slower than open discussions but is less likely to make people feel uncomfortable. (It also prevents *elite capture* of the discussions.)

Inconvenience nonetheless caused

The most common inconvenience I am aware of having caused was the inconvenience of waiting. Part of this was the result of poor planning and bad roads, and part of it was the consequence of (attempted) staging. People were not treated in a rural clinic until I had arrived and seen how long the queue was (in order to be impressed, I assume, by the extent of the demand). Children stood outside their school, long before my actual arrival, to sing me a welcome song. More commonly, people were asked to be at a venue well before I could possibly arrive. In at least two cases, organisations attempted to stage my visits, even though I had not disclosed which villages I would be visiting until the morning of the visits, by alerting *all* relevant people in *all* villages covered by a programme of my potential visit. In these cases, this led to many people waiting for the foreigner who never arrived.

Accountability towards key informants

Accountability is the obligation to explain, justify and take responsibility for one's actions. Researchers in the social sciences are accountable to their financers, supervisors, peers and wider readership, as well as, importantly, to their research participants. The May 2018 version of the Dutch "Code of ethics for research in the social and behavioural sciences involving human participants" does not use the term *accountability*. However, it does cover the need to explain and justify one's actions, in the form of requirements in relation to 'informed consent', which is the requirement that respondents are aware of the context in which the engagement takes place, and agree to be part of it.

My interview practice is largely compliant with this code of ethics' informed consent requirements (as well as everything else covered in this code), but there are a few exceptions. I note the following:

- The code of ethics says that "By default informed consent is active, i.e., through a deliberate act of the participant ('opt-in')".[14] Hardly any of the focus group discussions I conduct are based on 'opt-in'. I generally sample from potential respondents and invite them to an interview, rather than issue an open invitation and see who might be interested. In most cases, the advantages of this system outweigh the drawbacks. Specifically: an opt-in system would bias my pool of respondents to the more confident and often more powerful respondents, and this goes counter to the Leave No One Behind agenda that is part of many of my evaluations (and sense of ethics).
- I do not follow the code's informed consent requirements if this seems cumbersome and of limited use. Specifically:

  1 I rarely provide my or anybody else's contact details.
  2 I do not mention anything about the way I would store what people would tell me.
  3 Unless it is part of the protocol that the evaluation team has agreed on, I only mention that interviews are confidential if I

have reason to believe that this might reduce a respondent's hesitation.

4   I do not normally mention that I might re-use people's responses for the purpose of publications and instead I make sure potentially sensitive observations are not directly or indirectly traceable to the people who made them or to the groups they belong to.

5   In schools and many other settings, I do not provide information about the interview "sufficiently in advance" – irrespective of what the code of ethics might mean by this. Instead, I sample there and then, and proceed with the interviews almost immediately – though never without an introduction and a confirmation that respondents agree to being interviewed.

6   I do not "keep adequate records of when, how and from whom informed consent was obtained". Instead, I proceed with the interview immediately upon people consenting to me doing so, without any further documentation.

The code of ethics does not cover the issue of informing research participants of the findings and conclusions of research, though this is a reasonable part of researchers "taking responsibility for their actions" – one of the elements of accountability. In practice, I rarely offer to provide individual participants with an overview of findings, but I do always follow up where participants have requested it. Whenever I conduct a country assessment, I present my tentative findings to the subject of the evaluation (at least to the senior team and often to a larger group) on the last day of the assessment. A few days after my departure, I put these findings in writing and submit a back to office note.

Data utilisation

This book uses data that I gathered while conducting evaluations between 2010 and 2023. Re-using these data is an ethical thing to do because interviewing people is an imposition, and using these interviews to the maximum extent possible is better than discarding useful data and conducting yet more interviews instead. As respondents mentioned in this book are not identifiable, the re-use of their data does not have implications for their privacy.

The 41st standard of the Code[15] is to "avoid unnecessary reuse of previously published texts of which you were the author or co-author". PhD theses are exempted from this standard and this book is largely based on the thesis for which the Erasmus University in the Netherlands awarded me a mid-career PhD degree in June 2022.

Notes

1 KNAW *et al* (2018) *Netherlands code of conduct for research integrity*, Koninklijke Nederlandse Akademie van Wetenschappen. This code "respects the scope of international framework documents", the most significant of which is ALLEA (2017, revised edition) *European code of conduct for research integrity*, European Federation of Academies of Sciences and Humanities; but "On certain points, the [KNAW] Code presented here offers more specifics and details than the ALLEA code" (both quotations from page 8).

2 KNAW *et al* (2018) *Netherlands code of conduct for research integrity*, Koninklijke Nederlandse Akademie van Wetenschappen.

3 Deans of Social Sciences in the Netherlands (23 May 2018) *Code of ethics for research in the social and behavioural sciences involving human participants*, Deans of Social Sciences in the Netherlands.

4 This is not my invention: it is standard practice for all reviews conducted by the Independent Commission for Aid Impact (ICAI). I found this system useful and ever since 2016 I score every finding in every evaluation I conduct.

5 In the case of ICAI reviews the protocol for interview introduction was therefore slightly different: instead of assuring confidentiality "within the team", I would say that my interview notes "will not be shared outside of ICAI".

6 As of 2016 I add a one-sentence summary to each quotation, at its start, as this speeds up the analysis.

7 I do this in case of contracts with direct clients only. I cannot do this in cases where I work through a consultancy company, as there is no direct contractual link between the subject of the evaluation and me.

8 As outlined in Stroebe, W., Postmes, T. and Spears, R. (2012) "Scientific misconduct and the myth of self-correction in science", *Perspectives on Psychological Science*, volume 7, issue 6, pages 670–688, in the section titled "Increasing the chance of discovery" on pages 682–683.

9 'FAIR' stands for findability, accessibility, interoperability, and re-usability. See go-fair.org/fair-principles.

10 KNAW *et al* (2018) *Netherlands code of conduct for research integrity*, Koninklijke Nederlandse Akademie van Wetenschappen, page 17.

11 In the stimulated recall technique, you help a respondent to relive an experience step by step in order to, for example, gain insight in what led to a specific decision, problem, or solution.

12 When walking while talking, you do not have much eye contact and you can focus much of the conversation on practical issues that are well within the respondent's comfort zone (e.g., "so why tomatoes?"). This creates a comfortable context in which you can then casually cover the more contentious issues ("so why did almost none of these plots go to women?").

13 Using the nominal group technique means that you invite people to speak in turn, irrespective of their position in the group.

14 Deans of Social Sciences in the Netherlands (23 May 2018) *Code of ethics for research in the social and behavioural sciences involving human participants*, Deans of Social Sciences in the Netherlands, with the quotation from page 8, Section D, bullet point 12.
15 KNAW *et al* (2018) *Netherlands code of conduct for research integrity*, Koninklijke Nederlandse Akademie van Wetenschappen, page 17.

Index

For Product Safety Concerns and Information please contact our EU representative GPSR@taylorandfrancis.com Taylor & Francis Verlag GmbH, Kaufingerstraße 24, 80331 München, Germany

Printed and bound by CPI Group (UK) Ltd, Croydon, CR0 4YY

01/05/2025

01858360-0001